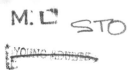

the Energy Crisis

opposing viewpoints

the Energy Crisis

opposing viewpoints

Bruno Leone
Judy Smith

OPPOSING VIEWPOINTS SERIES

Greenhaven Press

**577 SHOREVIEW PARK ROAD
ST. PAUL, MINNESOTA 55112**

©Copyright 1981 by Greenhaven Press, Inc.

ISBN 0-89908-303-X Paper Edition
ISBN 0-89908-328-5 Library Edition

CONGRESS SHALL MAKE NO LAW... ABRIDGING THE FREEDOM OF SPEECH, OR OF THE PRESS

first amendment to the U.S. Constitution

2200431

The basic foundation of our democracy is the first amendment guarantee of freedom of expression. The OPPOSING VIEW-POINTS SERIES is dedicated to the concept of this basic freedom and the idea that it is more important to practice it than to enshrine it.

TABLE OF CONTENTS   Page

the Opposing viewpoints series

THE IMPORTANCE OF EXAMINING OPPOSING VIEWPOINTS

The purpose of this book, and the Opposing Viewpoints Series as a whole, is to confront you with alternative points of view on complex and sensitive issues.

Perhaps the best way to inform yourself is to analyze the positions of those who are regarded as experts and well studied on the issues. It is important to consider every variety of opinion in an attempt to determine the truth. Opinions from the mainstream of society should be examined. Also important are opinions that are considered radical, reactionary, minority or stigmatized by some other uncomplimentary label. An important lesson of history is the fact that many unpopular and even despised opinions eventually gained widespread acceptance. The opinions of Socrates, Jesus and Galileo are good examples of this.

You will approach this book with opinions of your own on the issues debated within it. To have a good grasp of your own viewpoint you must understand the arguments of those with whom you disagree. It is said that those who do not completely understand their adversary's point of view do not fully understand their own.

Perhaps the most persuasive case for considering opposing viewpoints has been presented by John Stuart Mill in his work *On Liberty*. Consider the following statements of his when studying controversial issues.

9

THE OPINIONS OF OTHERS

If all mankind minus one were of one opinion, and only one person were of the contrary opinion, mankind would be no more justified in silencing that one person than he, if he had the power, would be justified in silencing mankind....

We can never be sure that the opinion we are endeavoring to stifle is a false opinion...

All silencing of discussion is an assumption of infallibility....

Ages are no more infallible than individuals; every age having held many opinions which subsequent ages have deemed not only false but absurd; and it is as certain that many opinions now general will be rejected by future ages....

The only way in which a human being can make some approach to knowing the whole of a subject, is by hearing what can be said about it by persons of every variety of opinion, and studying all modes in which it can be looked at by every character of mind. No wise man ever acquired his wisdom in any mode but this....

The beliefs which we have most warrant for have no safeguard to rest on but a standing invitation to the whole world to prove them unfounded....

To call any proposition certain, while there is any one who would deny its certainty if permitted, but who is not permitted, is to assume that we ourselves and those who agree with us are the judges of certainty, and judges without hearing the other side....

Men are not more zealous for truth than they are for error, and a sufficient application of legal or even social penalties will generally succeed in stopping the propagation of either....

However unwilling a person who has a strong opinion may admit the possibility that his opinion may be false, he ought to be moved by the consideration that, however true it may be, if it is not fully, frequently, and fearlessly discussed, it will be a dead dogma, not a living truth.

From *On Liberty* by John Stuart Mill.

A pitfall to avoid in considering alternative points of view is that of regarding your own point of view as being merely common sense and the most rational stance, and the point of view of others as being only opinion and naturally wrong. It may be that the opinion of others is correct and that yours is in error.

Another pitfall to avoid is that of closing your mind to the opinions of those whose views differ from yours. The best way to approach a dialogue is to make your primary purpose that of understanding the mind and arguments of the other person and not that of enlightening him or her with your solutions. One learns more by listening than by speaking.

It is my hope that after reading this book you will have a deeper understanding of the issues debated and will appreciate the complexity of even seemingly simple issues when good and honest people disagree. This awareness is particularly important in a democratic society such as ours, where people enter into public debate to determine the common good. People with whom you disagree should not be regarded as enemies, but rather as friends who suggest a different path to a common goal.

ANALYZING SOURCES OF INFORMATION

The Opposing Viewpoints Series uses diverse sources; magazines, journals, books, newspapers, statements and position papers from a wide range of individuals and organizations. These sources help in the development of a mindset that is open to the consideration of a variety of opinions.

The format of the Opposing Viewpoints Series should help you answer the following questions.

1. *Are you aware that three of the most popular weekly news magazines, Time, Newsweek, and U.S. News and World Report are not totally objective accounts of the news?*
2. **Do you know there is no such thing as a completely objective author, book, newspaper or magazine?**
3. **Do you think that because a magazine or newspaper article is unsigned it is always a statement of facts rather than opinions?**
4. **How can you determine the point of view of newspapers and magazines?**
5. **When you read do you question an author's frame of reference (political persuasion, training, and life experience)?**

Many people finish their formal education unable to cope with these basic questions. They have little chance to understand the social forces and issues surrounding them. Some fall easy victims to demagogues preaching solutions to problems by scapegoating minorities with conspiratorial and paranoid explanations of complex social issues.

I do not want to imply that anything is wrong with authors and publications that have a political slant or bias. All authors have a frame of reference. Readers should understand this. You should also understand that almost all writers have a point of view. An important skill in reading is to be able to locate and identify a point of view. This series gives you practice in both.

DEVELOPING BASIC THINKING SKILLS

A number of basic skills for critical thinking are practiced in the discussion activities that appear throughout the books in the series. Some of the skills are:

Locating a Point of View The ability to determine which side of an issue an author supports.

Evaluating Sources of Information The ability to choose from among alternative sources the most reliable and accurate source in relation to a given subject.

Distinguishing Between Primary and Secondary Sources The ability to understand the important distinction between sources which are primary (original or eyewitness accounts) and those which are secondary (historically removed from, and based on, primary sources).

Separating Fact from Opinion The ability to make the basic distinction between factual statements (those which can be demonstrated or verified empirically) and statements of opinion (those which are beliefs or attitudes that cannot be proved).

Distinguishing Between Prejudice and Reason The ability to differentiate between statements of prejudice (unfavorable, preconceived judgments based on feelings instead of reason) and statements of reason (conclusions that can be clearly and logically explained or justified).

Identifying Stereotypes The ability to identify oversimplified, exaggerated descriptions (favorable or unfavorable) about people and insulting statements about racial, religious or national groups, based upon misinformation or lack of information.

Recognizing Ethnocentrism The ability to recognize attitudes or opinions that express the view that one's own race, culture, or group is inherently superior, or those attitudes that judge another race, culture, or group in terms of one's own.

It is important to consider opposing viewpoints. It is equally important to be able to critically analyze those viewpoints. The discussion activities in this book will give you practice in mastering these thinking skills.

Using this book, and others in the series, will help you develop critical thinking skills. These skills should improve your ability to better understand what you read. You should be better able to separate fact from opinion, reason from rhetoric. You should become a better consumer of information in our media-centered culture.

A VALUES ORIENTATION

Throughout the Opposing Viewpoints Series you are presented conflicting values. A good example is *American Foreign Policy*. The first chapter debates whether foreign policy should be based on the same kind of moral principles that individuals use in guiding their personal actions, or instead be based primarily on doing what best advances national interests, regardless of moral implications.

The series does not advocate a particular set of values. Quite the contrary! The very nature of the series leaves it to you, the reader, to formulate the values orientation that you find most suitable. My purpose, as editor of the series, is to see that this is made possible by offering a wide range of viewpoints which are fairly presented.

David L. Bender
Opposing Viewpoints Series Editor

THE ULTIMATE SOLUTION?

THE ULTIMATE SOLUTION

Dick Wright. *The Providence Journal-Bulletin.* Reprinted with permission.

On October 17, 1973, the Organization of Petroleum Exporting Countries (OPEC) created a shock wave when they abruptly shut off their oil pipelines to the industrialized nations of the world. That action was shortly followed by a second stunning blow when they more than doubled the price of crude oil. The signs of an impending energy crisis had surfaced on previous occasions, but never, before the OPEC actions, had the world been affected on such a wide scale and in so ominous a way.

15

Dick Wright's cartoon, "The Ultimate Solution," is a tongue-in-cheek answer to the problems of imported oil dependence, rising energy costs and lack of a consistent national energy policy. Several of the viewpoints in this anthology deal with the problem of alternative energy development. For example, Barry Commoner's argument supporting increased use of solar power (Chapter 4, Viewpoint 3) typifies Wright's "soft energy" or "renewable resources" approach. Conversely, Samuel McCracken (Chapter 4, Viewpoint 4) attacks not only solar technology per se, but also claims that dependence on such technology would adversely effect the social structure of the nation.

However the real question is to decide if, indeed, there is an energy crisis. To consumers and many energy experts who are being wedged between ever rising energy costs and dire predictions of energy shortages, the answer seems to be an unqualified yes. On the other side, many authorities in the business, academic, government and scientific communities deny that such a crisis exists. Despite inflated costs and seeming shortages, it has been urged that the crisis is artificial and brought on by a wide range of possible causes from unnecessary governmental regulation to underhandedness by profit hungry energy industries.

This anthology of Opposing Viewpoints attempts to present both sides of the energy issue in a fair and balanced manner. The issues debated range from the validity of the crisis itself to possible energy alternatives. It is therefore the editors' hope that this book will assist the reader in arriving at an informed opinion on a problem whose significance is affecting and will continue to effect the future of humankind.

Chapter 1

Is There An Energy Crisis?

1

"Today the U.S. and all the world's importers of oil are caught in an acute and lasting energy emergency."

There Is An Energy Crisis

Harold Chucker

Harold Chucker is associate editor of the *Minneapolis Star* and specializes in foreign affairs and economics. A former Ford Foundation Fellow in economics at Columbia University, he has received several journalism awards for articles dealing with the economy. Mr. Chucker is a member of the Society of American Business Writers and the American Economic Association. In the following viewpoint, he presents a grim update on the energy situation and concludes that the current situation demands immediate accelerated conservation.

Consider the following questions while reading:
1. **What is the attitude of America's Western allies toward the rate of energy consumption in the U.S.?**
2. **What do authors Stobaugh and Yergin believe is our best solution for the energy crisis?.**
3. **Do you agree with their assessment? Why or why not?**

Harold Chucker, "Energy Situation Turns to Gloomy," *The Minneapolis Star,* February 12, 1980. Reprinted by permission from *The Minneapolis Star.*

Time has been telescoped. The emergency we were told was 20 years off has arrived.

That's the grim "update" of the energy situation made by the co-editors of "Energy Future," the best-seller published last year. The authors of the update, contained in a year-end edition of Foreign Affairs magazine, are Harvard Professors Robert Stobaugh and Daniel Yergin.

The evidence that 20 years of anticipated change have been telescoped into about 18 months—mid-1978 to the end of 1979—lies in these events:

From $12-$13 per barrel in late 1978, oil prices surged to the $30-$35 range, a level that forecasters had not anticipated until the year 2000. In absolute terms, Stobaugh and Yergin say, the price increases are greater than those of 1973-74.

Political threats to the world's oil supply, discussed as potentially serious five to 10 years in the future, became critical last year.

OPEC PRICE-SETTING

The price-setting mechanism of the Organization of Petroleum Exporting Countries has broken down. The cartel became unstable last year, with its members engaged in a pricing free-for-all, leapfrogging each other in setting ever higher prices. Power within OPEC has tilted toward the price hawks who have no concern about the effects of their actions on the world's economy.

The international energy system, designed to allocate adequate supplies among the consuming nations, "seems to be slipping out of any rational control." The Western European nations, in particular, are increasingly sniping at the U.S. for Americans' failure to curb their appetite for oil.

The U.S. is becoming a captive of "hostile oil." A major share of our imported oil is under the actual or prospective control of governments that "regard the West as the enemy."

Because of the telescoping of time, Stobaugh and Yergin fear, "the difficulty of meeting the challenge is much greater than would have been the case had its gravity become apparent" after the 1973-74 oil embargo shock: "Responses that might have been sufficient between 1974 and 1979 no

19

longer suffice; today the U.S. and all the world's importers of oil are caught in an acute and lasting energy emergency."

How we suddenly got this way, after what the two professors call "a kind of Indian summer for world oil" — the apparent calm and stability in the 1974-78 period — gets detailed treatment in the Foreign Affairs article.

PRICE CONTROLS

Price controls shielded Americans from the realities, Stobaugh and Yergin say. President Carter's "constructive and sensible" energy program of April 1977 was sneered at, and a "prescient and pessimistic" Central Intelligence Agency study of the world oil market was dismissed by critics. The critics, claiming there is a lot of oil in the world, and that OPEC would produce it rapidly to maximize profits, encouraged complacency.

SERIOUSNESS OF THE SITUATION

First and most obviously, the energy problem is pervasive and acute. It reminds me of the cable once sent to Whitehall by a traveling diplomat, who said, "Impossible to exaggerate seriousness of situation here, but I will try."

It is no exaggeration to say that the energy problem is central to the economic well-being and national security of all industrialized nations. Access to energy supplies is likely to be the dominant foreign policy issue of the 1980's.

E. G. Jefferson, President, E. I. duPont de Nemours & Company, *Vital Speeches of the Day*, July 1, 1980.

As the doubts about an oil crisis grew, a weariness over the warnings of a threat set in. And with that weariness came a refusal to restrain demand, to give up the habits of the oil junkie.

When Iran, in the chaos of its revolution, temporarily cut off supply, the U.S. officially blamed OPEC for the long gasoline lines. It was all a matter of supply, we were told. But the Europeans, increasingly snappish, put a good share of the blame on U.S. gluttony. While they had cut their consumption of oil, they pointed out, the U.S. had increased its demand. In their view, "the potential of conservation had barely been touched" in this country.

"Our priorities are askew," Stobaugh and Yergin write. They note that last year $1.5 billion was spent on existing federal conservation programs. So far this year the U.S. bill for imported oil is running at the rate of $80 billion a year...

The professors see some encouraging changes. They find a growing consensus that America's "most important energy source for the 1980s could be conservation of energy." The decontrol of oil prices, they feel, will help in encouraging greater energy efficiency—conservation.

The prospects for any sharp cuts in oil imports remain "uncertain," however. We may, in fact, be lulled into further complacency as the economy slows, a possible oil glut develops and prices ease, or slow their rate of increase. But given the political instability of many of the OPEC nations, and the chances for new disruptions of supply, that's dangerous thinking!

What about alternative sources of energy? Only coal can be counted on for additional supply over the next decade, but that may be accompanied by a decrease in the combined output from domestic oil, gas and nuclear sources. As for alternative sources—synthetic fuels, solar power, biomass, windmills, etc.—no substantial help is expected in this decade, even with massive subsidies.

CONSERVATION

A slow movement toward greater conservation won't do. Stobaugh and Yergin caution. That would mean closing down plants because of inadequate energy, along with higher unemployment, colder houses and, for some, a choice between food and fuel. The authors call for an accelerated program of conservation, supported by large incentives—40 to 60 percent of the investment for better insulation, for instance, plus long-term financing for the rest.

The aim, the authors say, should be zero energy growth for this decade. "Meeting this goal through productive conservation is the best way to promote positive economic growth." It

21

"So long as that idiot chooses to stand there...can you blame me?"

can be done, they claim, noting that some large companies are setting and meeting negative energy growth targets— getting along with less energy without curbing production.

Except to those who put their faith in consipracy theories and will never believe there is an energy problem, it should be clear that events in the international oil market have turned a bleak energy future into a gloomy energy present. The emergency that exists must be taken seriously, and must be accompanied by the kind of belt-tightening that Americans have so far shunned.

"It should be apparent that our problem is a matter, not of natural shortage, but of unnatural government regulation."

There Is No Energy Crisis

M. Stanton Evans

M. Stanton Evans has been a contributing editor to *Human Events* since 1968. A graduate of Yale University, he has also served on the editorial staff of *The Freeman* and *National Review* magazines. Mr. Evans has authored several books including *Revolt on Campus* (1961), *The Liberal Establishment* (1965) and *Assassination of Joe McCarthy* (1970). In the following viewpoint, he scoffs at predictions of doom and insists that America has sufficient amounts of natural gas and oil reserves. It is "unnatural government regulation," he contends, which is largely responsible for the energy situation.

Consider the following questions while reading:
1. **What predictions did the U.S. Geological Survey make regarding the supply of U.S. oil?**
2. **What is meant by "proved reserves"?**
3. **What does Mr. Evans mean when he writes that our problem is one of "unnatural government regulation"?**

M. Stanton Evans, "The United States Has Plenty of Energy," *Human Events*, September 22, 1979. Reprinted by permission from *Human Events*.

Among the many fallacies enfolded in "the energy crisis," none is more durable than the notion that we are running out of fuel supplies, including oil and natural gas.

This idea has been repeatedly stressed by political spokesmen, and given a certain cachet of responsibility by scholarly-sounding no-growth advocates, such as the Club of Rome. To hear these people tell it, we are currently paying the price of over-consumption; having used up our energy resources with reckless abandon, we must learn to tighten our belts and do without...

Such arguments usually rely on data saying we have "proved reserves" amounting to only a 10-year supply of gas or oil, at present rates of consumption. As President Carter put it a couple of years ago, "we could use up all the proven reserves of oil in the entire world by the end of the next decade."

From which it is thought to follow that drastic measures on the energy/conservation front are going to be required.

All of this may sound quite plausible to those who haven't studied the energy record. But it simply isn't so. There is plenty of gas and oil out there awaiting us — far more than the figures reflected in the "proved reserves." The problem is one of going out and getting it.

RISING RESERVES

It is worth observing in such discussions that we have heard these cries of "wolf" before. In his new book, *An American Renaissance*, Rep. Jack Kemp (R.-N.Y.) recalls that "the U.S. Geological Survey has predicted the imminent exhaustion of U.S. oil almost on a regular basis — in 1914, 1926, 1939 and 1949. In 1936, for example, official data showed that if we continued to use petroleum at even the depressed 1933 rate, the country would run out of 'proved reserves' in 15 years. Yet U.S. proved reserves of petroleum kept rising — from seven billion [barrels] in 1920 to 19 billion in 1940 to around 30 billion in early 1978."

All these predictions of impending doom made the mistake that is being made today — confusing "proved" reserves with energy potential. As noted by the Independent Petroleum Association of America, proved reserves are a kind of ready inventory — the supply the producer has on hand to deal with current and foreseeable demand. They are "the quantity of oil which can be produced with today's equipment and technology from wells already drilled, given the price of crude oil at the time the estimate is made."

Since 1946, the IPAA points out, proved reserves of crude oil, while increasing in absolute size and keeping with consumption, have constantly hovered around the 10-year supply mark, decreasing only gradually in recent years as domestic production has fallen off. The largest inventory of reserves, measured on this basis, has been a 13-year supply in 1949–50; the lowest-recorded in 1978—slightly more than a nine-year supply.

M. Stanton Evans

SUPPLY AND DEMAND

The following table from the IPAA shows selected years in the past three decades, proved reserves in millions of barrels as estimated on a yearly basis, annual production in millions of barrels, and the number of years the proved reserves would last at that particular rate of production:

U.S. CRUDE OIL RESERVES*

Year	Proved Reserves	Production	Years' Supply
1948	20,874	1,726	12.1
1949	24,649	1,819	13.6
1954	29,561	2,257	13.1
1959	31,719	2,483	12.8
1964	30,991	2,644	11.7
1969	29,632	3,195	9.3
1974	34,250	3,043	11.3
1978	27,804	3,030	9.2

The amount of ready-inventory gas and oil, measured in this fashion is always far less than the potential supply. As demand intensifies, and prices are pushed up, producers realize there is a bigger market than they had previously thought, and increase their efforts accordingly. At the same time, the higher prices people are willing to pay in a time of upward demand pressure make the acquisition of fuel supplies through more expensive technology a feasible proposition.

As a result of this process, proved reserves — of oil or just about anything else — are almost always increasing in absolute terms, not decreasing. Herman Kahn and his associates at the Hudson Institute point out that known reserves of oil, iron, phosphates, bauxite, chromite, etc., all increased dramatically between 1950 and 1970 — by several hundred, and in some cases, several thousand, percentage points. Prof. Neil Jacoby of UCLA notes that proved world petroleum reserves increased nine-fold between 1948 and 1972, from 62.3 billion to 568.8 billion barrels.

CRISIS OF PRICE CONTROLS

If this is the way the process has worked before, why isn't it working now? The answer to this is plain enough: Through our system of price controls, we have prevented the proper signals from being transmitted to the energy producers, and denied to them the rate of return that would make it economic to bring forth additional petroleum through more expensive techniques.

*In millions of barrels

26

In this respect, it is worth observing that proved reserves of crude oil in the United States almost doubled in the period from 1946 to 1970, going from 21 billion to 39 billion barrels. In 1971, however, price controls were imposed, and proved reserves have steadily declined in each year since, dropping down to something less than 28 billion by the end of 1978. Without controls, that is, our domestic supply of crude oil was constantly increasing; with controls, it has been just as steadily decreasing.

NO ENERGY SHORTAGE

I would like to stress again — as I have many times in the past — that the United States has no energy shortage! What we have had in the past, and still have, is over-regulation and a desperate shortage of reason in the processes by which energy and environment policies were drafted, adopted, and enforced.

Michael T. Halbouty, *Vital Speeches of the Day,* February 1, 1981.

From all of which it should be apparent that our problem is a matter, not of natural shortage, but of unnatural government regulation.

"In our energy crisis, we can't see the disaster dead ahead, and the time we must use to avoid it is fast running out."

Oil Reserves Are Shrinking

Americans for Energy Independence

Americans for Energy Independence (AFEI) is a nonprofit coalition of members from the business, labor, academic, scientific, industrial, consumer, conservation, ethnic and religious communities throughout the nation. An advocacy group, its purpose is to pursue effective energy policies which promote economic growth, social opportunity and national security. The following viewpoint is taken from an advertisement AFEI placed in *Reader's Digest* magazine. In it, AFEI attempts to outline the course which should be taken in order to achieve national energy independence.

Consider the following questions while reading:
1. **How does energy usage today compare with energy usage in the year 1890?**
2. **What energy measures does the author feel we must take in the immediate future, the intermediate future, and the future beyond the year 2000?**

"On the Titanic, They Didn't See the Iceberg Until Too Late," Americans for Energy Independence, 1976. Reprinted by permission.

Like passengers on the Titanic, we Americans are dancing while we head for what could be a disaster in energy. True, we saw small "ice floes" in the long gas lines and heating fuel shortages a few years ago. But most of us didn't—and don't yet—realize that the "floes" meant deadly "icebergs" ahead. We don't understand how much time it will take to put our country on a safe energy course. We could, like the men on the bridge of the Titanic, end up spinning the wheel loo late...

The U.S. relies on oil (and natural gas) for some 75% of its energy. With the economy improving, we have imported oil to the point of near flood. Foreign oil now supplies almost half our demand. Within two years, it may hit two thirds.

Any day—tomorrow—the organized oil exporting countries of the Mideast could stop most of the flow, as they did in 1973, and cripple us in a matter of days. We live extremely vulnerable, but very few Americans know it.

But even if the exporting countries didn't block the flow, eventually—and relatively soon—it will stop anyway. For the world simply will not have enough oil to keep burning it as a fuel. Though we don't see it, we already have a crisis situation. It will get worse every year. And from two directions: As oil and natural gas reserves dwindle, we and the rest of the world keep using more and more, faster and faster.

OXEN FOR TRACTORS

In the next 25 years, the U.S. will consume about as much energy as in all of its history. Today, the average American uses the power of 110 horses (compared to less than one in 1890). And we will need more, because some Americans still don't have their fair share of energy.

Neither do hundreds of millions of people in other parts of the world, many of whom are just now trading oxen for tractors and claiming other machine-made comforts of modern living. On earth today, we count four billion energy consuming people. By the year 2000, we will have six billion.

What happens if the world can't meet the energy needs of all these people? A precipitous drop in standards of living everywhere, and the higher the standard, the greater the drop. In countries like ours—so dependent upon energy— even a small shortage can bring great suffering. As Congressman Mike McCormack of Washington has said, by 1985, being short a million barrels of oil a day—only about 2%—

29

Let's learn to live with the energy, not the problems

LePelley in *The Christian Science Monitor* © 1980 TCSPS.

will mean a loss of some 900,000 American jobs.

HOW MUCH TIME?

Do we have enough time left to salvage the situation? Barely, when we consider what engineers call "lead time"— the time between deciding what to do and actually getting it done. We have precious little lead time left to do what we must in energy. What's more, delay in using our lead time can bring much worse than delay in achieving objectives. What we can do, say, in 10 years if we start now, might take 12 years if we start next year. Or, if we wait too long, *we may find it too impractical to do at all!*

Dividing the future into three periods, we must use our remaining lead time to achieve these objectives in each period:

The Immediate Future, 1976-1985: Call this a breath-holding period. Because we must still rely mostly on oil and natural gas. Will the oil exporting countries let us buy what we need from them? Will we pump enough of our own—and conserve enough of *all* we have, both imported and domestic —to make it?

This is also the lead time period in which we must move vigorously toward higher coal production and also more nuclear power.

Though coal represents America's most abundant fossil fuel, we need a long lead time to get at it the right way. We can't just go back to old-time coal mining. Deep mines must be increasingly mechanized and made safer. We must "scrape" much of our coal from near the ground surface, and re-beautify the land afterward. We must burn the coal according to new technologies that reduce sulphur emissions to harmless levels. We'll need new and expensive manufacturing plants to convert coal into synthetic gas and oil to extend our use of these fuels...

The Intermediate Future, 1985-2000: Coal and nuclear power will supply most of our energy, but *only if* we have already done the big job on coal and *only if* we make full use of our nuclear option. We'll still be burning some oil and natural gas. And a small percentage of our energy will start to come from such sources as the sun.

In this period, nuclear power will have to carry a continually larger burden of generating electricity. It can do so, but *only if* we once again remain conscious of lead time. Today, some 20 years after the first commercial nuclear plant at Shippingport, Pa., 59 plants provide 9% of our electricity. It takes about 10 years to build a nuclear plant, and, according to a Bureau of Mines study, we'll need 900 of them to carry the load by 2000...

The Future Beyond 2000: Assuming we've taken the right course, we could be in safe waters. Our coal resources could last several centuries at least. New nuclear reactors called "breeders" actually make more fuel than they burn. And, in 25 to 50 years, some of those promising energy sources of the future—solar, wind, tides, geothermal, biomass, fusion—should be working to make electricity on a large scale.

A GLOOMY FORECAST

A U.S. Department of Energy study predicts that American oil production will decline 14.4 percent to 8.1 million barrels a day by 1985, even under the best circumstances. The report says oil output could rebound to 9.3 million barrels daily by 1995, mainly as a result of new discoveries in Alaska.

Some oil-company studies are even gloomier. An Exxon Corporation study predicts that by the year 2000, U.S. oil production will decline to around 8 million barrels a day while natural-gas output will dip to 14.1 trillion cubic feet a year.

Daniel Yergin and Robert Stobaugh of Harvard Business School conclude in Energy Future, *a study of the nation's energy problems, that U.S. oil production will fall to 6.5 million barrels daily by 1990.*

"Americans should not delude themselves into thinking that there is some huge hidden reservoir of domestic oil that will free them from the heavy cost of imported oil," the report warns.

U.S. News & World Report, May 4, 1981.

At the moment, solar energy seems most promising for direct heating of rooms and water. Using it to generate electricity is another matter. As yet, we don't have the technology to build a practical demonstration plant. Which means we really have no hard facts on what to expect from solar or other future sources in the next century—only high hopes, if we use our lead time for all the research and development still to be done.

"There is, in the free world today, a 36-year supply of proven reserves already staked out and producible at today's prices."

Oil Reserves Are High

Yale Brozen

Yale Brozen is a professor of Business Economics at the Graduate School of Business, University of Chicago and an Adjunct Scholar of the American Enterprise Institute for Public Policy Research. He has been on the faculty at the University of Chicago since 1957. The author of several economics textbooks, Dr. Brozen has also written *Advertising and Society* (1974) and *The Competitive Economy* (1975). In the following viewpoint, he outlines several energy "myths" which he claims have been created by the federal government.

Consider the following questions while reading:
1. **List the seven myths regarding oil that Prof. Brozen has outlined.**
2. **According to the author, what are some of the reasons we seem to be experiencing oil shortages?**
3. **Do you agree with his arguments? Why or why not?**

Yale Brozen, "The Mythology of Energy," *The Freeman,* July, 1979.

The war against the automobile and against private enterprise continues. This time, it appears in the guise of a quest for a reduced international payments imbalance and freedom from coercion by the Organization of Petroleum Exporting Countries. Propaganda almost as crude and just as untruthful as that used by the Allies in World War I is the major instrument in the current MEOW (Moral Equivalent of War) campaign for expansion of taxation and government power.

The campaign uses several myths in its attempt to sell Americans on ceding more of their freedom to the central government. Here is a list of the more blatant falsehoods accepted and propagated by the opinion manufacturing establishment.

1. The world will run out of oil in the 1980s.
2. The severe international payments imbalance is caused by the high usage and high price of imported oil.
3. An oil-rooted adverse payments balance is causing the dollar to depreciate, causing import prices in dollars to rise and, as a consequence, causing inflation.
4. We are vulnerable to an oil embargo by the Mid-East countries.
5. The gasoline shortage and long lines at filling stations in late 1973-early 1974 were caused by the oil embargo in effect at that time.
6. We must reduce our vulnerability to an embargo by accumulating a one-billion-barrel stockpile of oil and by cutting energy usage.
7. The government must plough billions into government-directed energy research to save us from ourselves and from foreign powers...

MYTH NUMBER ONE

Let us take the myths and examine each. Myth number one is that the world will run out of oil in the 1980s. Actually, it is unlikely that we will run out of oil by the 2080s. There is, in the free world today, a 36-year supply of proven reserves already staked out and producible at today's prices.

The number of years' supply of proven reserves is at the highest level in the history of the statistic. Traditionally, proven reserves have ranged from fifteen to thirty years at contemporaneous rates of oil use. Moreover, the statistic is only indirectly related to the actual amount of oil existing underground in the world, and even the direction of the relationship is unclear, because exhaustion of prospects produces

34

a rise in price, and hence makes previously worthless reserves worth "proving."

How much more oil remains to be discovered that is producible at today's prices is unknown. Geologists' estimates range from a low of a twenty-year additional supply to a high of fifty years.

Taking the lowest estimate, today's real prices need not change for the coming half century to induce a supply of petroleum sufficient to meet all demands. At prices 50 percent higher than today, producible reserves in sight more than double. It would become worthwhile to use the enormous shale oil deposits in Colorado, Utah, and Wyoming. Of

Yale Brozen

35

the 1.87 trillion barrels of oil in shale, 600 billion are recoverable at the higher price. That is enough to supply us for another 100 years. There are also staggering reserves available in the Canadian Athabasca tar sands and the Missouri, Kansas and Oklahoma tar sands which would become economically workable at the higher price...

MYTHS NUMBER TWO AND THREE

If oil imports cause an adverse balance of payments or if the great increase in crude oil prices in 1974 were a cause of an adverse balance of payments, then Germany and Japan should be in much deeper trouble than we. They import *all* of their crude oil while we import less than half. They import all of their natural gas while we import only a small fraction. Yet their balance of payments is positive. While the dollar declined, the mark and the yen appreciated. The cause of the payments imbalance and the decline of the dollar is the string of unprecedented peacetime federal deficits since 1973...

MYTHS NUMBER FOUR AND FIVE

Why did we have those long lines at gasoline stations in 1974? Was it because of the Arab embargo?

The reason for those long lines was because the Federal Energy Office allocated gasoline and gave orders to refiners as to what products they could produce. *All during the period of the embargo, our stocks of gasoline, crude oil, and other petroleum products in storage kept increasing.* Crude oil was still being imported. Instead of coming from the Mid-East, it came from Canada, Indonesia, Venezuela, and Nigeria. Some came indirectly from Libya and other Mid-East countries via Curacao and the Bahamas.

The embargo made only a small difference in the volume of imports. The oil companies did a massive and heroic job redirecting world trade. Routing of oil was changed in some cases and sources in other cases. But the Federal Energy Office screwed up the works. It underallocated gasoline to metropolitan areas, such as Chicago, New York, and Washington, and it overallocated to rural areas. City residents wasted gasoline by driving far into rural areas to fill their tanks.

Are we subject to possible blackmail by embargo? The answer is a clear no! During the Arab embargo, we imported from other sources and indirectly from the Mid-East countries that were embargoing us. Libya knew its oil was coming to us, but as long as it was labeled as going elsewhere when it left Libyan ports, Libya was glad to get the revenues.

There are more alternative sources available today than there were in 1974. Mexico is now supplying us with growing amounts. Venezuela has 20 percent of its capacity shut down and available. Nigeria is a bigger producer now than it was in 1974. Dome Petroleum is starting full scale development and transportation out of the Canadian Arctic. China is now exporting oil.

REGULATIONS PREVENT PRODUCTION OF DOMESTIC RESERVES

Sometimes you get the feeling the country is going stark, raving mad.

America's growing dependence on imported oil is eroding the value of the dollar, skewing the balance of trade, crippling the domestic economy and feeding unemployment.

The problem, astonishingly enough, is not that we cannot produce virtually all the oil we need domestically. The problem — at least a major part of it — is the rules and regulations that keep us from exploiting our vast domestic resources.

William Raspberry, *Washington Post*, 1980.

MYTH NUMBER SIX

We are now developing storage facilities and accumulating a one-billion barrel stockpile of oil, at a cost of $25,000,000,000 purportedly to make ourselves less vulnerable to any future embargo. The Arabs must be laughing themselves sick all the way to the bank as we turn over $15,000,000,000 to them for oil we are going to stick back in the ground (in old hollowed out salt domes).

Is it really necessary to accumulate a stockpile to reduce

our vulnerability to an embargo? The answer is no! Many countries are willing to supply us if the Arabs cut us off, including some Arab countries if we cover up the fact that they are supplying us...

MYTH NUMBER SEVEN

Finally, we come to the myth that the government must plough billions of dollars into energy research if the new technology is to be developed to provide the energy we need when oil runs out in the 1980s. First, let's recognize that a shortage is a business opportunity. If anything in demand is likely to run short, its price will rise. Anyone developing a substitute or an additional supply will find plenty of eager customers.

With the increase in the price of home heating fuels, suppliers began offering automatic damper controls which cut the use of fuel by 20 percent. When fuels were cheap, it was not economic to install automatic damper controls; they could not pay for themselves. The capital it would have taken to produce them was more productive in producing gas than in saving gas. Production of the controls would have been a waste of metal, plastic and workers' time. These factors of production were conserved by the more efficient expenditure of capital on gas discovery and production.

As it became increasingly costly to produce gas, capital began to flow into damper controls where it could save more gas than it could produce. The investment now pays for itself.

The rise in the price of energy is inducing the production of energy saving equipment and of less energy intensive motors, engines, generators, cement kilns, furnaces, boilers, refrigerators, freezers, air conditioners, and water heaters. It is also attracting investment into private Research and Development (R&D) to develop alternative sources of energy, to develop techniques for secondary and tertiary recovery of oil from spent fields, and to improve methods of extracting oil from shale and tar sands. In 1975, oil companies invested $51 million in coal R&D, $38 million in developing methods for converting coal into synthetic fuels, $30 million in oil shale R&D, $9 million in tar sands R&D, $7 million in geothermal R&D, and $2 million in solar R&D...

The private market does a superior job in allocating resources to their most productive uses, including choosing among alternative R&D programs, than the government does.

If the government wouldn't try to do so much, we would get more accomplished, and energy would be more plentiful than it is now.

"The nation has enormous natural gas resources currently available, with the prospects that even larger amounts will come on stream in the next few years."

Natural Gas: The Alternative to OPEC

Fred J. Cook

Fred J. Cook has been a freelance writer since 1959. Prior to that time, he was editor of the *New Jersey Courier* and a feature writer for the *New York World Telegram and Sun.* Mr. Cook is the author of several books including *The FBI Nobody Knows* (1964) and *What So Proudly We Hailed* (1968). In the following viewpoint, he urges greater use of natural gas as the solution to the energy crisis claiming that it is the cleanest, cheapest and most abundant fuel resource in the nation.

Consider the following questions while reading:

1. **What were the findings of Dr. Vincent McElvay and Dr. Paul H. Jones?**
2. **What would some of the potential oil savings be through the greater use of natural gas?**
3. **If Mr. Cook's information regarding the abundance of natural gas is correct, why, in your opinion, is the U.S. government (and others) claiming that an energy crisis is upon us?**

"Abundant Energy, The Natural Gas Boom," July 12, 1980. *The Nation Magazine,* Nation Associates Incorporated © 1980.

Natural gas reserves are so huge that they amount to "about ten times the energy value of all [previous] oil, gas and coal reserves in the United States combined," according to Dr. Vincent McElvay, director of the U.S. Geological Survey, in a speech in Boston in 1977. Dr. McElvay was reflecting on the findings of a geological survey which concluded that zones of seabed along the Louisiana and Texas coasts alone held 24,000 trillion cubic feet of gas—the equivalent of 4 trillion barrels of oil, roughly twice the estimate of the ultimate world resources in petroleum.

NATURAL GAS RESERVES

According to Dr. Paul H. Jones in 1978: "There's good scientific evidence that this brine [in the Tuscaloosa Trend along the Louisiana–Texas coasts] could contain as much as 50,000 trillion cubic feet of [natural] gas. That's equal to 2,500 times our yearly production." Dr. Jones, a veteran geologist widely respected as the foremost authority on the area, later changed this estimate to 100,000 trillion cubic feet. "From all I hear now, my 50,000 trillion prediction looks conservative," he said in May 1980.

"The Greater Anadarko Basin [a broad strip stretching across lower Oklahoma and the Texas Panhandle) ranks right up there with the Gulf Coast for its gas resources," said Bill Dutcher, associate of Robert Heffner III, pioneer wildcatter of the Anadarko and chairman of the American Gas Association's Independent Gas Producers' Committee. Heffner has predicted that deep-drilling in the Anadarko (not counting a veritable pincushion of shallow, conventional wells) could yield some 60 trillion to 300 trillion cubic feet of gas.

"This is the major uncovered story in the world. Anyone who has seriously gone after gas has found it. All sorts of new fields are being opened up, with 20 million to 30 million cubic feet a day coming on stream," said a respected researcher who spent more than a year tracking down the natural gas story.

PRODUCT OF PROPAGANDA

The man is right. Serious, large-scale drilling of deep natural gas wells did not begin until 1979, and, in a matter of months, new fields with seemingly limitless potential have been developed. Estimates now considered ultraconservative hold that we have sufficient natural gas supplies to last for at least 100 years. The U.S. Geological Survey's assessment of the reserves of the Gulf Coast alone would mean that we have potential resources to last for 1,150 years.

What, then, becomes of the much–trumpeted energy "crisis"? It is clearly the product of propaganda. While there are undeniably serious problems, these have been blown out of all proportion by scare techniques...

There are a number of solutions available to an America willing to use all its natural and technological resources, but the quickest and best is the solution that has been ignored in all of established wisdom's scripts—natural gas. It is the cleanest, cheapest and most abundant fuel resource in the nation...

NATURAL GAS SHORTAGE WAS MANMADE

The present natural gas shortage is not due to nature's stinginess; it is manmade. Its proximate cause has been the Federal Power Commission's (FPC) enforcement of low ceiling prices on all interstate sales of natural gas. Because they have been too low, these price ceilings have discouraged firms from exploring for and developing new natural gas reserves.

Richard B. Mancke, *Energy and the Way We Live,* 1980.

Consider the potential savings gained through greater use of natural gas. The Department of Energy's figures for 1979 show that residential and commercial uses of petroleum, mainly for heating and cooking, totaled 29.526 quads (a quad represents 1 trillion B.T.U.s, or British Thermal Units, of energy). Each quad eats up some 653,163 barrels of crude oil. In other words, these 29.526 quads represent approximately 19,285,290 barrels of oil. The industrial sector consumes another 28.791 quads—or an additional 18,805,215 barrels of oil. A grand total of 38,090,505 barrels. As the Hudson Institute has pointed out, "Oil or gas used as industrial fuels can easily be switched... Thus there is much room today to use more gas and less oil, or vice versa... Moreover, gas can be converted to liquid fuels [for transportation, which consumes another 19,777 quads] if necessary..."

USES OF NATURAL GAS

The potential of natural gas is obvious. Granting that a complete switch from oil to gas is out of the question, any large-scale alteration of usage would save literally millions of barrels of oil. Major pipelines deliver gas to most sections of the nation; in residential, commercial and industrial areas, street-line delivery systems were installed long years ago. Thousands of homeowners converted to natural gas last year because gas heat costs about half as much as oil. The switching has been limited in some sections by temporary shortages of gas and by a demand so great that suppliers literally ran out of gas boilers and conversion units. Common sense says that, with abundant supplies now coming on line and with furnace units available, further large-scale conversions should be encouraged. If even half of those residential-commercial-industrial quads could be saved, there would be some 19 million barrels of oil we would not have to import...

A GAS RICH NATION

As should be clear from this account, the nation has enormous natural gas resources currently available, with the prospects that even larger amounts will come on stream in the next few years as the indubitably rich sediments of areas like the Tuscaloosa Trend, the Anadarko and the Overthrust Belt are more fully probed. There are immense amounts of natural gas that can flow to the lower forty-eight states from Alaska's North Slope as soon as pipeline connections can be made. This is the certainty. More speculative, but not to be ignored, are other possibilities for the future. Geologists, using new seismic techniques, believe that an Overthrust Belt similar to the Rockies' runs along the entire length of the Appalachians. They know that the waters under Lake Erie contain large gas deposits because Canadians have drilled wells on their side of the lake. The American side has remained untouched because of controversies over environmental issues. The evidence mounts that productive gas wells have been discovered in deep-drilling off the New Jersey coast. Although the companies involved have kept a tight lid on information, a spokesman for the New Jersey Department of Energy recently estimated that construction of a gas pipeline to bring this new bounty ashore might be started in the next six months.

Developments of the past two years have clearly vindicated Doctors McElvay, Knudsen and Jones, the pioneering experts who tried to tell the nation we have so much natural gas that we do not have to remain prisoners of OPEC. All of the evidence indicates that massive shifts in the use of gas in the commercial–residential–industrial sectors of the economy could be made; even that some electric plants could be fueled with gas instead of imported, high–priced oil. Natural gas, obviously, is the quickest and best alternative to OPEC.

6

"North America will become essentially self-sufficient in oil, with hardly any imports required from overseas."

Energy Conservation: The Alternative To OPEC

S. Fred Singer

S. Fred Singer is currently a professor of Environmental Sciences at the University of Virginia, Charlottesville. He is also a member of the Energy Policy Studies Center at the University. In the following viewpoint, Dr. Singer claims that reliance upon OPEC for oil has diminished considerably since American and world oil consumers have adopted energy conservation.

Consider the following questions while reading:
1. **According to Dr. Singer, what information does the average American get concerning the gasoline situation, and why does he consider that information incorrect?**
2. **Explain what fundamental change the author describes that has taken place in the world oil market?**
3. **What are the far-reaching consequences he foresees from the reduction in oil use?**

S. Fred Singer, "Hope for the Energy Shortage," *Newsweek*, May 18, 1981. Reprinted by permission of the author.

To the average American, the energy problem is mainly his monthly fuel bill and the cost of filling up his gas tank. He may also remember that in 1979, and way back in 1974, he had to wait in long lines at gasoline stations. For all of this, he blames the "Arabs" or the oil companies or the government, or perhaps all three. Much of the information that he gets from the media, as well as his own past experience, tells him that energy prices will continue to go up sharply and that gas lines are going to come back whenever a conflict flares up in the Middle East.

But this view of the future is not correct—notwithstanding the conventional wisdom. Oil prices have more or less stabilized, and may even fall. Even oil embargoes should no longer be feared. When Iranian mobs occupied our Teheran embassy in November 1979, Jimmy Carter declared that we would stop buying Iranian oil—really a self-imposed embargo. If Khomeini had made the same statement, there probably would have been a panic here. As it turned out, no one even noticed the embargo. Other imports took the place of Iranian oil, and life went on calmly for the most part. And with domestic oil prices set free, the allocation of gasoline will take care of itself and prevent long lines, should a sudden shortage develop.

ENERGY CONSERVATION IS WORKING

Be this as it may, the U.S. consumer has reacted to the high price of oil and to his own expectations, and has followed the example of industry by adopting energy conservation. Home insulation, passive solar houses, and a more careful use of heating and air conditioning have become popular. The motorist has fallen in love with small, fuel-efficient cars. He had that kind of love affair before, back in 1975. But when Congress gave him the wrong signals and told him that gasoline prices were going to stay put, the trusting consumer went back to gas guzzlers. This time, the romance is for real, thanks to the 1979 price jump in oil. Not surprisingly, our imports dropped by 20 per cent in 1980 compared with 1979. In addition, the price of oil has now been deregulated by the Administration. The consumer is no longer treated to oil which has been made cheap at the expense of greater imports from overseas.

Neither consumers, nor politicians, nor indeed Arab oil sheiks seem to recognize that a fundamental change is taking place in the world oil market, right before our eyes. Oil consumption here and in most countries has been decreasing

46

sharply since a peak was reached about two years ago.

When the world oil price quadrupled in 1974, powerful economic forces were set into motion. When the price again doubled in 1979, these forces accelerated. They are now probably irreversible. All over the world, oil is being replaced wherever possible by cheaper fuels, mainly by coal and nuclear energy. Unlike Americans, foreign consumers use oil mainly as a boiler fuel (and hardly for transportation), so substitution is quite simple. These shifts, coupled with conservation, will cut world consumption of oil in half by the next decade.

IMPROVED ENERGY EFFICIENCY

The best energy investments available to the United States are in improved energy efficiency. Dollar for dollar, $250 billion dollars will save more energy if invested in conservation than it will produce if invested in any new sources of supply. However, no matter how much we conserve, we must develop new sources. Our entire society rests upon a diminishing fuel supply, and a transition must be made to more bountiful sources.

Denis A. Hayes, *Energy: Conservation and Public Policy,* 1979.

SWITCHING TO ALTERNATIVE FUELS

Here in the United States, only 11 per cent of all electric power is made from fuel oil. But many countries without coal, natural gas or hydropower must import expensive oil. For them, the decision to switch fuels is easy. Even Russia, an exporter of oil, is building nuclear plants just as rapidly as it can so it can export more oil and earn more hard currency. We are learning to make mixtures of ground–up coal and water which will burn in boilers designed for fuel oil. With proper attention to coal cleaning and pollution control, such substitution should have little effect on the environment. But it will save a great deal of money.

Conservation will have a large impact also on transportation. For example, General Motors tells us that in 1985 their

47

average passenger car will be getting 31 miles per gallon, more than the 27.5 mpg mandated by Federal law, and double the 1973 value. And if the urban minicar takes hold, gas mileage might jump to 60 mpg. General Motors even believes that there is a market for an electric car which would use no gasoline at all, but only electricity generated mostly by coal and nuclear energy.

S. Fred Singer

BECOMING SELF-SUFFICIENT

This dramatic cut in oil use will have far-reaching consequences. For one thing, North America will become essentially self-sufficient in oil, with hardly any imports required from overseas. Our outflow of dollars will be reduced and so will the income of OPEC, the Organization of Petroleum Exporting Countries. OPEC's oil production may shrink to very low levels by the next decade. This will cause tremendous upheavals in the Arab states, which have been enjoying multibillion dollar surpluses. It will be good news to oil consumers everywhere else. Our energy costs may not go down very much, but the fuels we use will be homegrown and more secure. The Middle East will become a much less interesting place.

By the next decade, old-timers will be trying to explain to youngsters that there was a President from the State of Georgia who believed that the energy problem was "the moral equivalent of war." When the youngsters ask "why?" I bet the old-timers won't remember.

DISTINGUISHING BETWEEN FACT AND OPINION

This discussion activity is designed to promote experimentation with one's ability to distinguish between fact and opinion. Consider the following example. If an ultra-liberal individual is elected president of the United States at some future date, that would be a fact. However, whether or not the United States would be better off socially, economically and politically with an ultra-liberal president is a matter of opinion. Future historians will agree that the individual in question was a president of the United States. But their interpretations of the political significance of his or her administration probably will vary greatly.

PART I

Instructions

Some of the following statements are taken from this chapter and some have other origins. Consider each statement carefully. Mark O for any statement you feel is an opinion or interpretation of facts. Mark F for any statement you believe is a fact. Then discuss and compare your judgments with those of other class members.

O = Opinion
F = Fact

_____ 1. Only coal can be counted on (as an alternative energy source) for additional supply over the next decade.

_____ 2. The energy problem is central to the economic well-being and national security of all industrial nations.

_____ 3. "I would like to stress again...that the U.S. has *no* energy shortage."

_____ 4. Our (energy) problem is a matter not of natural shortage, but of unnatural government regulation.

_____ 5. Access to energy supplies is likely to be a dominant foreign policy issue of the 1980's.

_____ 6. Gasoline prices in foreign countries have been higher than gasoline prices in the U.S. for many years.

_____ 7. A major share of American oil supplies is under the actual or prospective control of foreign governments.

_____ 8. America's growing dependence on imported oil is eroding the value of the dollar.

_____ 9. The best energy investments available to the United States are in improved energy efficiency.

_____ 10. Natural gas is the cleanest, cheapest and most abundant fuel resource in the United States.

_____ 11. Oil prices will probably fall in the upcoming years.

_____ 12. All over the world, oil is being replaced whenever possible by cheaper fuels such as coal and nuclear energy.

PART II

Instructions

STEP 1

The class should break into groups of four to six students.

STEP 2

Each small group should try to locate two statements of fact and two statements of opinion in this chapter.

STEP 3

Each group should choose a student to record its statements.

STEP 4

The class should discuss and compare the small groups' statements.

BIBLIOGRAPHY

The following list of periodical articles deals with the subject matter of this chapter.

Business Week — *Gloom Behind the Natural Gas Bubble,* April 23, 1979, p. 68.

S. Chapman — *Energy: The Myth of Independence,* **Atlantic,** January, 1981, p. 11.

Foreign Policy — *Energy Nightmares,* Fall, 1980, p. 132.

Deane R. Hinton — *Energy: Continuing Crisis,* **Department of State Bulletin,** February, 1981, p. 47.

David P. Hunt — *The Economics of Energy,* **The Freeman,** May, 1980, p. 278.

Z. Khalilzad and C. Benard — *No Quick Fix for a Permanent Crisis,* **Bulletin of the Atomic Scientist,** December, 1980, p. 15.

National Geographic — *Special Energy Issue,* February, 1981, p. 1.

Kenneth R. Sheets and Joseph Benham — *Oil Fever Rages — And So Does an Old Debate,* **U.S. News & World Report,** May 4, 1981, p. 49.

M. Sheils and W. J. Cook — *Two Gloomy Views of an Energy Future,* **Newsweek,** January 5, 1980, p. 49.

U.S. News & World Report — *No Shortage of Ideas to Solve Energy Crisis,* May 12, 1980, p. 70.

Washington Monthly — *Energy Crisis,* October, 1980, p. 10.

James A. Weber — *Energy Ethics: A Positive Response,* **The Freeman,** April, 1981, p. 195.

P.D. Zimmerman — *How to Beat the Energy Crisis,* **Newsweek,** April 28, 1980, p. 18.

Chapter

Is Nuclear Power An Acceptable Alternative?

1

"A nuclear accident with large-scale loss of life remains very improbable."

The Case For Nuclear Power

Petr Beckmann

Petr Beckmann came to the United States in 1963 from his native Czechoslovakia. He was educated at Prague Technical University and since 1965 has been professor of Electrical Engineering at the University of Colorado. He is the founder and editor of *Access to Energy,* a monthly newsletter, and the author of over 60 articles published in scientific journals. His books include *The Health Hazards of Not Going Nuclear* (1976). In the following viewpoint, Mr. Beckmann insists that the danger of loss of life is greater from other fuel sources than from nuclear power and focuses upon what he considers the real reason behind opposition to nuclear power.

Consider the following questions while reading:
1. **List several of the arguments the author offers in support of nuclear energy.**
2. **Why, according to Mr. Beckmann, is the disposal of nuclear wastes safer than that of coal and other fossil fuels?**
3. **Compare Mr. Beckmann's arguments favoring nuclear energy with those of Peter Fong.**

Petr Beckmann, "Nuclear Power: The Safest Available," *The Daily Telegraph,* London, England, August 7, 1979. Reprinted by permission from the publisher.

The case for nuclear power is one of simple morality: It saves most of the lives now lost in generating comparable amounts of electricity from less safe sources. If it were not also reliable, economic and assured of free world supplies for millennia, we would be faced with a moral dilemma, but fortunately, it is all of these as well.

The comparison with our present (or suggested future) power sources is one that any layman can check out for himself. Moreover, the results have rarely been disputed by the nuclear foes; they have simply been ignored. And the comparison works not just for safety from accidents; it works equally well for radiation, waste disposal and terrorism.

Of course nuclear power is not perfectly safe—no large-scale power source can be—but per unit of energy produced, its price in human suffering is much smaller than for any other source. Nor can this be disputed by pointing to the short history of nuclear power. Not only do we have several reactor-centuries of experience, but in the 22-year history of nuclear power, many thousands have died in the fossil fuel cycle—even when only the correspondingly small fraction of energy is considered.

NUCLEAR EXPLOSION?

A nuclear power plant cannot undergo a nuclear explosion; the only danger is a significant release of radioactivity, and that danger is localized in a few cubic meters of space, where it can be surrounded by a multilayered defense in depth. Moreover, the time scale of a possible accident is so slow (melting of the fuel, melting through the pressure vessel, possible failure of the containment building) that there is time to bolster the defenses wherever they are in danger of growing weak. And even if this slowly progressing battle threatens to be lost, there is time to evacuate the endangered area.

Nuclear safety, then, is not based on the infallibility of operators or the perfect function of gadgets, but on defense in depth and slow time scale. No other energy facility has even one of these two protections.

The dam of a hydroelectric plant, for example, can break and kill thousands in minutes, because it lacks a second, third and seventeenth safety dam, and because there is no time to take countermeasures.

THE HARRISBURG ACCIDENT

Both points were dramatically illustrated by none other than the Harrisburg accident. After four horrible failures, both human and mechanical, there was not a single death or injury. What other 845-megawatt power plant could contain such a sequence of failures without loss of life?

Yet at Harrisburg the defense in depth never retreated close to a meltdown; and even a meltdown would most probably have been contained without loss of life, for the prime purpose of the containment building is to contain the danger after a meltdown.

As for the time scale, one of the teams of experts called to the scene at Harrisburg had the Ralph Nader-like task of engaging in "what-if" fantasies to prepare for any possible further failures. They found all credible failures protected by backup systems, but if the entire electric supply failed, they had only one auxiliary diesel aggregate to generate emergency power, so they flew in a second one. What other 845-megawatt facility gives that kind of time when it threatens disaster?

Reprinted by permission from the *Manchester Union Leader*.

A nuclear accident with large–scale loss of life remains very improbable, but even if one were to happen, the loss of life could hardly be the same as we are now tolerating for other power sources. Some 20,000 Americans die premature deaths every year due to coal-fired power plants; throughout the world, coal-fired power takes a toll of between 40 and 200 lives per year per 1,000 megawatts (mostly via air pollution, but also in transportation and in the mines). Oil fires, oil smoke, gas explosions, not to speak of hydropower (2,000 deaths in a dam failure in Italy in 1963), can kill victims by the tens of thousands—and with incomparably higher probabilities.

Radiation? The natural radioactive background in Colorado is twice the American average (the difference from sea level is equivalent to 5,000 nuclear plants), yet the cancer rate is 30 percent below it...

WASTE DISPOSAL

Waste disposal is perhaps the biggest single advantage of nuclear power (which may explain why it has been singled out for such a ferocious attack). Compared to coal and other fossil fuels, it has two overwhelming advantages: The volume is miniscule, and the toxicity temporary. Nuclear wastes are the only wastes of an industrial society that can be completely and permanently removed from the biosphere.

A 1,000–megawatt nuclear unit produces 2 cubic meters of wastes per year; a coal-fired plant produces 20 pounds of solid wastes not per year, but per second. Nuclear wastes can be solidified, sealed into glass, put into waterproof, earth-quake–proof steel drums, buried in salt formations (which will seal up rather than let water enter) and monitored for 650 years, by which time their radioactivity will decay below the level of the ore they originally came from.

Coal wastes, on the other hand, contain 19 toxic metals (arsenic is one). They will be around forever, and some of them are "disposed of" in your lungs.

NUCLEAR TERRORISM

Can nuclear power be abused by terrorists? Yes, but not very effectively. They can inflict incomparably heavier losses much more easily by other means (which I am not about to describe here). Some of the scares are plain silly. Plutonium dispersal, for example. What self-respecting terrorists would use a weapon that kills 15 to 40 years later, would not kill many people even then, is extraordinarily hard to acquire, can immediately be detected in absurdly small quantities, and is more valuable than gold?

Reliability? In America 10.6 percent of electric capacity is nuclear, but it pulls almost 13 percent of the load. And that in spite of the fact that the government will shut down not one but all nuclear plants of the same type as soon as a ludicrously small fault is detected or even merely suspected.

NUCLEAR POWER AND THE QUESTION OF RISKS

Radiation and the risks of radiation command considerable public attention. However, it is not generally realized that safety regulations are much stricter for radioactive materials than for other dangerous substances. For example, nuclear power stations emit radioactive materials; oil-and-coal-fired power stations discharge sulphur dioxide...

The emission limits for nuclear power stations are 100 times lower than they are for oil- or coal-fired stations... Furthermore, in the case of coal, it has been estimated that in Pennsylvania 30,000 miners died in the mines between 1870 and 1950—an average of about one man a day for 80 years. Next to such appalling tolls, the safety history of the nuclear power industry is uniquely encouraging.

Radiation—A Fact of Life, American Nuclear Society.

Then there is availability. The already mined and stored American supplies of uranium ore fertile for breeding could supply America with a century of electricity. So why the opposition?

That is not an easy question. Perhaps much of the "why?" is answered by the "who?"

Who, indeed? Statistically speaking (with plenty of

Petr Beckmann

exceptions), the well educated, the affluent, the nonproducers, the ones who do not make a living by customers coming to them voluntarily but who are used to planning, analyzing and redistributing at other people's risk.

They are not against nuclear power in particular, but against an abundance of energy in general, against industrial and economic growth, and most of all against capitalism, the system that has brought material wealth and political freedom to more people than any other in history.

"Nuclear power stations...have turned out to be much more costly than conventional generating plants."

The Case Against Nuclear Power

Anthony Lewis

Anthony Lewis is a columnist for *The New York Times* and is currently Lecturer on Law at Harvard Law School. A two time Pulitzer Prize winner, Mr. Lewis has been associated with the *Times* since 1948 except for three years as a reporter for the *Washington Daily News.* His books include *Gideon's Trumpet* (1964) and *Portrait of a Decade* (1964). In the following viewpoint, Mr. Lewis reviews an article which challenges the assumption that nuclear power is desirable and concludes that the economic and human cost of nuclear power is forbidding.

Consider the following questions while reading;
1. **What effects, according to the author, would nuclear power have upon oil shortages?**
2. **Why does Mr. Lewis believe that the use of nuclear power plants would contribute to the spread of nuclear weapons?**

Anthony Lewis, "The New Case Against Nuclear Power," the *Minneapolis Tribune,* June 20, 1980. © 1980 by The New York Times Company. Reprinted by permission.

The arguments for and against nuclear power have come to seem at the same time familiar to us and numbingly difficult to resolve. But the terms of the debate may be changing. New circumstances, new facts, put the old issues in a different light—and greatly strengthen the case against nuclear plants to generate electricity.

PRACTICAL ARGUMENTS

That is the message of a provocative article in the summer issue of the quarterly *Foreign Affairs,* just out. It is a disturbing piece, one that means to shake assumptions on a fundamental subject and does. Its scientific points will be debated by the experts. But much of its logic rests on practical arguments that ought to be meaningful to politicians and ordinary citizens in a democracy.

The piece, called "Nuclear Power and Nuclear Bombs," is by Amory B. Lovins, a physicist and consultant on energy policy; his wife, L. Hunter Lovins, a lawyer, and Leonard Ross, a former California public-utility commissioner who now teaches law at the University of California, Berkeley. Amory Lovins published a ground-breaking article on the "soft path" in energy—the development of renewable resources—in *Foreign Affairs* four years ago. That became a basic reference point in the energy debate, and the new piece may well have the same kind of impact.

We have assumed, the article says, that the worldwide spread of nuclear power is economically desirable, is necessary to reduce dependence on oil and can be regulated by international agreement so that it will not lead to the proliferation of nuclear weapons. The authors squarely challenge those assumptions.

HIGH COST NUCLEAR

Nuclear power stations, they say, have turned out to be much more costly than conventional generating plants. Between 1971 and 1978, one study shows, the capital cost per kilowatt went up twice as fast for nuclear as for coal plants, even including the need for the latter to meet rigorous anti-pollution standards. Nuclear now costs 50 percent more than coal, and tighter safety regulation after Three Mile Island will increase the differential.

That economic reality is reflected in the market. All over the world, plans for installation of nuclear generating plants have been cut way back. The United States, Brazil, West Germany, China—all kinds of societies have simply found the economic cost too forbidding...

NOT OIL SUBSTITUTE

Nor, they say, is nuclear power a rational substitute for oil. Nuclear plants make electricity, and only about a tenth of the world's oil goes for that purpose. Most of it is used for such things as vehicle fuel and petrochemical feedstocks.

So the most massive increases in nuclear power would have little effect on the urgent questions of world oil prices and supplies. For example, quadrupling Japan's nuclear capacity by the year 1990 would reduce its dependence on imported oil by about 10 percent.

PLAYING RUSSIAN ROULETTE

American technology, the government and the corporate giants of the nuclear industry are playing Russian roulette with the health and lives of American citizens. The potential for disaster far outweighs the meager benefits society derives from nuclear energy. Given the state of the art of American technology and based on past performance, a case can be made for the inevitability of a nuclear disaster; it came within an eyelash of occurring at Three Mile Island.

Don Winters, *Minneapolis Star*, Aug. 7, 1980.

"Most governments," the article says, "have viewed the energy problem as simply how to supply more energy of any type, from any source, at any price, to replace oil—as if demand were homogeneous."

But in fact the price and nature of nuclear power makes it economically viable for only about 4 percent of all energy needs.

NUCLEAR WEAPONS

As to proliferation of nuclear weapons, the article makes

some new and extremely worrying technical arguments. It challenges what has been a premise of all international efforts to keep nuclear weapons from spreading to countries that do not now have them: that power reactors can be designed, operated and monitored so that they do not produce material of practical use in making bombs.

All present power reactors produce, as waste, what is called "reactor-grade plutonium," which for various reasons has been considered as impractical as material for bombs. But in

Anthony Lewis

fact, the authors say, governments or "some subnational groups" could make it into bombs as good as those now made from "weapons-grade plutonium"—or, alternatively, power reactors could be operated so as to produce the latter without greatly increasing costs or being detected.

"We cannot have nuclear power without proliferation," they conclude, "because safeguards cannot succeed either in principle or in practice." But ending the nuclear power program would make it possible to limit the spread of weapons and detect breaches of international controls, because goods and services now used for both reactors and bombs would then be "unambiguously military in intent."

Is it "a fantastic, unrealistic, unachievable goal" to wind down nuclear power programs? No, the authors say; governments would just have to obey the economic principles to which they profess allegiance. They would just have to stop applying "heroic measures to resuscitate and artificially sustain the victim of an incurable attack of market forces."

"To abandon nuclear power," they say, "does not require any government to embrace anti-nuclear sentiment or rhetoric. It can love nuclear power—provided it loves the market more."

"Once the aura of mystery is stripped from it, nuclear hazard is no more different from the other hazards of life, such as auto crashes and coal mine explosions."

Nuclear Power Is Acceptable

Peter Fong

Peter Fong, a native of China, is currently a professor of physics at Emory University in Atlanta. He has also taught at the Massachusetts Institute of Technology, the California Institute of Technology and Cornell University. He is the author of several technical textbooks including *Energy and Our Environment* (1976). In the following viewpoint, Dr. Fong attempts to explain why nuclear power has created such a controversy and then warns that only those nations which fully utilize nuclear energy "will be the future masters of the earth."

Consider the following questions while reading:
1. **According to Dr. Fong, why do most people fear nuclear power? Do you agree or disagree with him? Why?**
2. **What does the author claim is a beneficial aspect of radiation?**

Peter Fong, "Radiation and Nuclear Risks." Reprinted from *USA Today*, July, 1980. Copyright 1980 by Society for the Advancement of Education.

Last year was eventful for nuclear power. The Three Mile Island incident in March jolted the world. It was followed in May by a massive rally in Washington against atomic energy. For a time, it appeared that nuclear power was finished...

The Three Mile Island incident has been referred to as the watershed. Many people, cautiously endorsing nuclear energy, beat a hasty retreat. What really happened?

During a tour after the incident, one Congressman, awed by the colossal structure and complicated machinery, sighed and commented that nuclear power is beyond his grasp and, as such, he would rather not have it. This is the most ridiculous impression that could come across to an intelligent person.

Whatever mysteries of nuclear physics that generate this energy have nothing to do with the accident. What is relevant to the accident is no more than the system of plumbing, which is the easiest thing everybody can understand. In comparison, what makes an airplane fly and crash is far more complicated and involved. Looking at it logically, it would take Herculean courage to take a seat in an airplane.

The essence of Three Mile Island is just a failure of the plumbing—something like a sewer backup. Just the same, it is messy and dirty, but is there anyone abandoning his home because of a sewer backup?...

THE MYSTERY OF NUCLEAR RISK

The crippled plant did release radiation, but the total effect of that is calculated to be at most one fatal cancer over and above the thousands normally expected in the area. Yet, the accident was painted as an unprecedented catastrophe. Why is a nuclear accident so fearsome?

Perhaps the feeling of the general public can be typically represented by ABC-TV newsman Howard K. Smith's comments. He said automobile accidents kill 50,000 people a year, but we accept it as a part of life. His wife, in a recent hospitalization, received eight X-ray shots, twice as much radiation exposure as that received on Three Mile Island. No one has yet been hurt by this nuclear accident, but a nuclear accident is different from all other hazards—coal mine explosions, airplane crashes, etc. The unseen radiation can cause incredible damage, including cancer many years later.

Reprinted by permission of Charles G. Brooks, *The Birmingham News.*

Thus, it is not that the nuclear hazard is great — there are far more devastating and far more frequent hazards. It is not that the hazard is not acceptable — we accept much worse. The problem is that nuclear radiation is mysterious, sneaky, insidious, inscrutable, and devilish.

This is not the first time that the human race has been faced with such a shockingly uncertain threat. For exactly the same reason, ancient people feared eclipses, lightning, and even sex, separating them from the more understandable dangers of lions, tigers, and bears. Human sacrifices were offered to placate the heavenly monster that gobbled up the sun — and sure enough, after the offering, the monster burped out the old Sol. Since then, even the most rational were convinced that heads must roll at eclipses...

NORMAL RADIATION

If nuclear radiation is so devilish, consider that every human body contains billions of radioactive atoms (potassium–40), each one of which is capable of emitting deadly, cancer-causing radiation. In fact, each person receives 24 millirem radiation per year from his own body — which constitutes one-fourth of the natural background radiation. Indeed, about 1,000 cancer deaths per year can be traced to this source — the victim's own body. Nuclear radiation is not a snake that spoils the otherwise tranquil and beautiful Garden of Eden. Every human body, in terms of radiation effects, is already as dirty as a low-level nuclear waste heap — and this is God's design without the help of nuclear power or atomic bombs (they do not involve potassium). Like aging and death, nuclear radiation is just a part of normal life, no more mysterious and inscrutable than mother's milk.

Everybody now knows that radiation can cause cancer. Does everyone know that radiation can cure cancer? Does everyone know that nuclear radiation is now used in many hospitals for this purpose and millions of lives have already been saved by it? With our indulgence in anthromorphic projection and with the guidance of the mere animal reflex, such a helping agent should at least be recognized as a familiar, friendly ally, albeit a hot one that must be handled with care. Who could be so vile and so ungrateful as to call such a benefactor sneaky, insidious, and devilish?

Once the aura of mystery is stripped from it, nuclear hazard is no more different from the other hazards of life, such as auto crashes and coal mine explosions. In terms of numbers of deaths per year, its total effect is so slight that slipping in a bathtub would be a more serious hazard in comparison...

NUCLEAR DOMINANCE

On nuclear power, the Russians and their allies not only plan expansion, adding 40 nuclear plants in the near future, but also intend to build nuclear power plants in their population centers, obviously to take advantage of the technology of "cogeneration," by which the total energy value derived from a nuclear plant would be tripled. They not only consider a nuclear plant to be safe, but safe enough to have it next door. The minute risk they are willing to take is a small price for the big prize of world domination. Primitive man was willing to take the risk of using fire, which is a dangerous form of energy and still kills 6,700 people a year, to become the dominant species on the earth...

DANGERS FROM NUCLEAR ARE PRESUMPTIVE

As for the intrinsic hazards to life of nuclear energy, with one exception — the persistence of radioactivity — they are no different, and in some respects are much smaller, than the intrinsic hazards of other energy-producing systems. As for actually killing people, the numbers seem to favor nuclear by a large margin. If we compare nuclear reactors with automobiles, the comparison is striking. The dangers from autos are real and stark: 50,000 people will die on the road every year. By contrast, the dangers from nuclear energy are presumptive — no one has been killed by radioactivity from a commercial nuclear power reactor.

Alvin M. Weinberg, *Bulletin of the Atomic Scientists*, April, 1977.

In the struggle for dominance, there is no choice of hard or soft technology. Either you win or you perish. The one nation that can fully utilize the most powerful energy source — nuclear energy — will be the future master of the earth. History is merciless to those who got off the train of progress and watched it speed by with folded hands.

"Fission power, a menace to our health, damages our physical system and our ecosystem in ways we hardly understand; at the same time, it is acually damaging to our social system as well."

Nuclear Power Is Not Acceptable

Mark Reader

Mark Reader has been an associate professor of political science at Arizona State University since 1967. He was educated at the University of Michigan and is editor of *Energy: The Human Dimension* (1977). In the following viewpoint, Dr. Reader asserts that nuclear power is destructive not only to people, but also to the environment, and that if we continue to depend on nuclear energy, we shall leave behind a legacy of problems for future generations.

Consider the following questions while reading:
1. According to Dr. Reader, what is the relationship between the "wartime atom" and the "peaceful atom"?
2. According to the author, how can we develop an energy policy that will establish peace?
3. What does the author point to as the major factor pushing us in the direction of a nuclear future?

Mark Reader, "Energy and Global Ethics." This article first appeared in *The Humanist* July/August 1979 and is reprinted by permission.

If we agree that to err is human, must we forego all energy technologies that, as they malfunction, would create havoc? I am thinking of nuclear accidents and of potential failures in the operation of high-technology solar space-stations, should they orbit the earth.

One is reminded of Murphy's Law: If things can go wrong, they will go wrong, precisely because humans are imperfect and because all objects carry with them the imprints of their makers and users...

A PEACEFUL ATOM

Of all energy sources, nuclear power is maximally destructive in terms of peace, equality, liberty, health, and harmony between and among people and their environments. In my judgment, there's no such thing as the peaceful atom. And nuclear power is that energy source which is maximally in disharmony, or discontinuity, with physical and social systems, as the recent accident at Three Mile Island indicates.

The individual nuclear reactor is really part of a system. What we have to do is to see nuclear reactors as part of the nuclear fuel cycle and the nuclear fuel cycle as the building blocks of a new culture, a culture which, in my view, cannot survive for very long without destroying those people who have built it and without destroying vast portions of the ecosystem.

When we buy nuclear reactors, we buy a way of life, a set of values, a set of furniture that we have to serve even as it serves us. And as we serve it, we do so in a situation of slavery rather than a situation of equality and freedom. It sets up the problem of taking care of the flow of radioactive material virtually forever and the need for surveillance, for security, for watching.

If we want to watch radioactive wastes, say through the half-life of plutonium—twenty-four thousand years—we are saying that we have to shape a civilization that can watch these wastes for periods of time as long as the last Ice Age.

If we choose to police our radioactive wastes (we have no idea of how to dispose of them), what we do is to deed to the future a civilization fundamentally like our own. It is doubtful that such a civilization can endure for the period of time necessary in order to watch these materials.

72

A PEOPLE'S PROBLEM

One reason the nuclear cycle is proliferating globally is because of the inequality of energy reserves, wealth, and power. The desire of the have-not countries to catch up with the haves, and the desire of the haves to dominate the have-nots, is propelling the world in the direction of a nuclear civilization. So what we are talking about is not simply an American problem or a Japanese problem, or a French or Iranian problem. It is a people's problem. And the mechanism that's fueling the proliferation of this cycle is inequality, and with it a potential for violence.

NUCLEAR WASTE: A FUTURE PLAGUE

Even if every nuclear plant in America were shut down tomorrow, we'd still face the terrible question of what to do with ten million cubic feet of high-level waste materials that already exist. Poisons that threaten our health and our lives, and will continue to plague future generations for thousands of years.

Dr. Peter Montague, Executive Director, National Campaign For Radioactive Waste Safety.

As the nuclear cycle spreads globally, what is also happening is that the number and volume of radioactive transactions is increasing. The number of points that have to be watched and taken care of is growing exponentially. Dr. Carl Johnson has estimated the amount of radioactive materials that will be produced at stationary atomic sources in the United States within the foreseeable future:

We face a geometric increase in contamination of the environment by nuclear plant emissions. These are radioactive noble gases, radioiodine, tritium carbon 14, radiocesium and plutonium and other transuranics. The environmental burden of tritium will increase by over one thousand times in the next forty years. Krypton 85 will increase from one hundred billion curies released in 1980 to nearly ten

73

trillion in 2020, and iodine 129 will increase from about one hundred curies in 1980 to over five thousand curies by 2020. For carbon 14, we expect an increase of four hundred times in the next twenty years. The release of plutonium and other alpha–emitting transuranics will increase by over one hundred times by the year 2020.

The nuclear fuel cycle, the culmination of Western culture generally, is not only placing a bind on the present, but it will force future generations to organize to contain this material.

PRESSURES ON FREEDOM

We have the problem of avoiding nuclear war. Within the next seven years as many as forty nations will have nuclear capability sprung from their atomic energy programs, and the pressure on human freedom will increase accordingly. Human freedom will become more problematic as the fuel cycle spreads. There is no reason to believe that our experience with the so–called peaceful atom will be any different from our experience with the wartime atom. And what we understand is that the wartime atom set up the need for the national security state. What the peaceful atom will do—or what radioactivity in international and national traffic will do—is increase that need everywhere.

There will continue to be growing instability, inequality among nations as a source of conflict. There will be "have" nations, or a limited technological elite which possesses the skills to use the atom, and those who resent their rule.

So the kind of world social system that nuclear power requires is fundamentally an inegalitarian social system. And insofar as there is inegalitarianism in the world, in terms of power or in terms of wealth or status, the pressure toward social change and social disruption will always be present.

THE PARADOX OF NUCLEAR POWER

This I call the nuclear paradox. What nuclear power requires is total social stability, but because of its fundamental inequality what it guarantees is that social stability cannot be achieved.

Fission power, a menace to our health, damages our physical system and our ecosystem in ways we hardly understand; at the same time, it is actually damaging to our social system as well.

Reprinted with permission from the *Minneapolis Tribune*.

It seems to me we must ask, then, how can we develop an energy policy that will establish peace? What can the human community do?

First, I think it is imperative that one of the major nuclear powers declare a moratorium on the development of all things nuclear—weapons and power plants. Once such a moratorium is announced, we must find ways to guarantee energy to every nation and spare them any recourse to the nuclear alternative. At the same time we should speed up the redistribution of the world's wealth, power and status. Like it or not, the major factor pushing us in the direction of an unhappy and intolerable nuclear future is the problem of continuing global inequalities.

We must look to the genius of government and the genius of people to figure a way to solve the energy crisis globally without resorting to a nuclear future. We must agree on some long-term goals.

DISTINGUISHING BETWEEN STATEMENTS THAT ARE PROVABLE AND THOSE THAT ARE NOT

From various sources of information we are constantly confronted with statements and generalizations about social and moral problems. In order to think clearly about these problems, it is useful if one can make a basic distinction between statements for which evidence can be found, and other statements which cannot be verified because evidence is not available, or the issue is so controversial that it cannot be definitely proved. Students should constantly be aware that social studies texts and other sources often contain statements of a controversial nature. The following exercise is designed to allow you to experiment with statements that are provable and those that are not. Some of the statements are taken from this chapter and some have other origins.

Instructions

In each of the following statements indicate whether you believe it is provable (P), too controversial to be proved to everyone's satisfaction (C), or unprovable because of the lack of evidence (U). Compare and discuss your results with your classmates.

P = **Provable**
C = **Too Controversial**
U = **Unprovable**

_____ 1. We cannot have nuclear power without proliferation because safeguards cannot succeed either in principle or in practice.

_____ 2. The worldwide spread of nuclear power is economically desirable.

_____ 3. Nuclear power stations have turned out to be much more costly than conventional generating plants.

_____ 4. Solid nuclear wastes can be safely buried in salt formations.

_____ 5. Radiation can cause cancer.

_____ 6. Once the aura of mystery is stripped from it, nuclear hazard is no more different from other hazards of life, such as auto crashes and coal mine explosions.

_____ 7. In the struggle for dominance, there is no choice of hard or soft technology. Either you win or you perish.

_____ 8. Of all energy sources, nuclear power is maximally destructive.

_____ 9. When we buy nuclear reactors, we buy a way of life, a set of values, a set of furniture that we have to serve even as it serves us.

_____ 10. Within the next (few) years as many as forty nations will have nuclear capability sprung from their atomic energy programs.

BIBLIOGRAPHY

The following list of periodical articles deals with the subject matter of this chapter.

Ralph Kinney Bennett — *Nuclear Power in Perspective,* **Reader's Digest,** June, 1981, p. 131.

J. George Butler — *Christian Ethics and Nuclear Power,* **The Christian Century,** April 18, 1979, p. 438.

Bernard L. Cohen — *King Coal and the Melt-Down Myth,* **National Review,** June 12, 1981, p. 667.

G. Reiger — *On a Nuclear Day, You Can Die Forever,* **Field and Stream,** December, 1979, p. 23.

Virginia Southhard — *The Ethics of Nuclear Power,* **America,** April 18, 1981, p. 321.

E.J. Sternglass — *Invisible Death,* **Nation,** February 28, 1981, p. 255; March 7, 1981, p. 267.

Marvin Stone — *A Case for Atomic Power,* **U.S. News & World Report,** October 29, 1979, p. 92.

Marcia Terry — *Nuclear Power "Convert" Attacks Some Common Fears,* **Human Events,** April 25, 1981, p. 19.

J. Viorst — *Is Nuclear Power the Only Choice for the Future?,* **Redbook,** February, 1980, p. 50.

A.M. Weinberg — *Is Nuclear Energy Necessary?,* **Bulletin of the Atomic Scientist,** December, 1980, p. 58.

E. Winchester — *Nuclear Wastes,* **Sierra,** July, 1979, p. 46.

Chapter 3

Is the Energy Consumer Exploited?

"The (oil) companies, not OPEC, are the real cartel."

Oil Companies Control Energy Supplies

Edwin Rothschild

Edwin Rothschild is director of the Energy Action Educational Foundation. In the following viewpoint, Mr. Rothschild accuses the major international oil companies of controlling the world oil supply and of manipulating the market.

Consider the following questions while reading:
1. **How did Clifton C. Garvin, chairman of Exxon, explain his company's "fantastic profits in 1979?"**
2. **How did the oil companies respond to the glutted oil markets of 1977–1978?**
3. **According to Mr. Rothschild, how did the oil companies respond to the OPEC price increase of December, 1978?**

Edwin Rothschild, "Oil Prices: OPEC is Not 'Real Cartel'," *The Christian Science Monitor,* June 24, 1980. Reprinted by permission from *The Christian Science Monitor.* © 1980 The Christian Science Publishing Society. All rights reserved.

The belief that OPEC "sets" world oil prices and is powerful enough to impose its will on the rest of the world is as misleading as is the belief that the major international oil companies (Exxon, Shell, British Petroleum, Texaco, Standard Oil of California, Mobil, and Gulf) have so little power that they have no choice but to accept OPEC's decisions. It comes as no surprise, therefore, that Clifton C. Garvin, chairman of Exxon, recently told Forbes Magazine that the reason for the companies' fantastic profits in 1979 "was the tightening in world markets that enabled OPEC to double prices."

What Mr. Garvin did not say and what Energy Action has concluded in a recent thorough and well-documented report is that the seven major companies deliberately and concertedly began that "tightening in world markets" in 1978 — prior to the Iranian oil production shutdown.

OPEC NOT TO BLAME

This conclusion is of significance for the public and especially for its elected representatives, most of whom have refused to recognize the menacing power of the international companies and the deleterious effects of their actions on the supply and price of oil in world markets. The major companies, not OPEC, initiated the latest round of oil price increases. For its part, OPEC has responded to the price leadership shown by the companies. And such price leadership could not exist were it not for the still vast power these major companies have over the noncommunist world's supply of oil.

What has always disturbed the major companies is the specter, actual or potential, of a surplus of oil in the market. To avoid a surplus, which depresses prices and profits, the major companies have usually found ways, legal and illegal, to limit the availability of oil supplies.

Prior to the early 1970s, the seven major companies were able, to a large extent, to determine the availability of oil in the noncommunist world. Following the so-called "OPEC Revolution," the companies' power over oil production was reduced. Yet, even though OPEC captured a large area of decisionmaking, the companies, because of their marketing power and their financial and technical expertise, retained great control over the supply and price of oil in the world. This power was exercised and made painfully evident in 1978 and 1979.

GLUTTED OIL MARKETS

All throughout 1977 and the first six months of 1978, the world crude oil markets were glutted. Prices were actually declining in both real and nominal terms. OPEC members like Algeria, Libya, and Nigeria were competing with each other, cutting prices to attract their international major oil company customers to purchase crude oil. Thus, despite its cartel label, OPEC was thoroughly incapable of cutting production

'THIS LAND IS MY LAND, THIS LAND IS OUR LAND, FROM COAL DEPOSITS TO THE COPPER DIGGINGS; FROM NUCLEAR POWER TO OFF-SHORE DRILLING, THIS LAND BELONGS TO ME AND MINE.'

to eliminate or even reduce the surplus. On the contrary, many OPEC members were trying to increase production.

The international majors, however, aware of OPEC's inability to set production rates, were worried about eroding prices and profits. Having the motive, capability, and opportunity, they started to close ranks and eliminate the "oversupply" of oil early in 1978 — well before the "crisis in Iran" and OPEC's December, 1978, conference. Because of their ownership of and preferential access to large volumes of crude oil, their ownership and control over the largest and most economical oil tankers, pipelines, export refineries, terminals, storage facilities, and marketing outlets in the most strategically important markets, the major companies were able to reduce their stocks of refined products, cut back refinery operations, reduce imports into major consuming areas and decrease oil production from OPEC — all in the face of increasing consumption. As a result, the companies were able to drive up prices first for refined products and then for crude oil.

Beyond their ability to fine-tune their worldwide refining, transportation, and storage operations to restrict the supply of oil products, the major companies also retained a large degree of flexibility in determining the production levels for various OPEC producers. In the first six months of 1978 OPEC crude production was reduced by 450 million barrels compared to the first six months of 1977.

THE ARAMCO PARTNERS

The four Aramco partners — Exxon, Texaco, SoCal, and Mobil — reduced their purchases of oil from Saudi Arabia alone by just over 316 million barrels in this period. In fact, between January and August, 1978, Aramco shipped 20 percent less oil than in the year before. Moreover, when the Saudis directed Aramco to increase the amount of lower-quality oil production relative to the better-quality oil, the companies delayed the implementation of this policy, precisely at the time when demand for the better-quality oil was greatest. .

Similar flexible oil purchase arrangements exist between the companies and other OPEC members. Together with their other worldwide operations, these arrangements have allowed the major companies to make some of the most important supply and price decisions in the marketplace.

The effective reestablishment of the companies' power to control the market also derives from their ownership of new, large, and profitable sources of crude oil in Alaska and the North Sea. Exxon and BP together control 74 percent of Alaska's Prudhoe Bay field, while all the majors together dominate production in the North Sea. In early 1978 as oil from these sources was entering the market, the companies

EXXON WOULD LOVE TO OWN THE SUN

We must restore competition and motivation to energetically pursue the development of alternative energy sources. To do so we must force the oil companies to disgorge the solar energy companies, the coal fields, the uranium mines and the distilleries they have been absorbing for the past decade. Exxon would dearly love to own the sun.

Robert J. Henle, *America*, February 14, 1981.

at first lowered prices and forced competing oil from OPEC producers out of the market. This action not only made OPEC oil less expensive to the companies but assured profits for the Alaskan and North Sea oils. Once these supplies were flowing, however, a tightening of supplies would mean greater profits on both the reserves in the ground and the inventories purchased at the lower prices.

OPEC THE SCAPEGOAT

In December, 1978, the OPEC members did not so much agree to increase the world price of oil as ratify the increases already set by the companies. The companies welcomed the OPEC decision, because public attention and anger could be directed at OPEC and away from themselves. This was particularly useful since the companies continued to drive up prices after the OPEC meeting. North Sea producers raised prices in January (and in later months) and only in response did some OPEC countries raise prices above the levels set in December.

The companies, not OPEC, are the real cartel. The companies, not OPEC, tightened supplies in 1978 and 1979. The companies, not OPEC, still control the oil flowing into consuming countries. The companies' ability to manipulate the market was even noted by former Energy Secretary Schlesinger. When asked why he didn't use his power to force the companies to refine more oil in 1979, he replied that the companies could, in response, simply choose "to defer the importation of the oil and keep it on the high seas for a longer period of time."

"There was no credible evidence that gasoline shortages resulted from collusive conduct by oil companies."

Oil Companies Do Not Control Energy Supplies

Herbert Schmertz

Herbert Schmertz has been Vice-President of Public Affairs for the Mobil Oil Corporation since 1974. An attorney who earned his degree at Columbia University, his publications include *Takeover* (1980). In the following viewpoint, Mr. Schmertz defends the oil industry against critics' charges of collusion.

Consider the following questions while reading:
1. **According to the author, what are the factors which led to an oil shortage?**
2. **What does Mr. Schmertz claim is the goal of those who criticize the oil industry?**

Herbert Schmertz, "Deluding the Public About Oil," *The Christian Science Monitor,* August 26, 1980. Reprinted by permission from *The Christian Science Monitor.* © 1980 The Christian Science Publishing Society. All rights reserved.

For well over a year now, ever since US motorists encountered severe shortages of gasoline in the spring of 1979, the American oil industry has been the target of allegations that the industry, and more particularly the major international oil companies, conspired to create shortages and drive up oil prices after the suspension of oil exports from Iran.

ALLEGATIONS CONTINUE

These allegations continue to be echoed even though long since proven false in a plethora of hearings and forums, the most recent exoneration coming in two reports issued by the US Department of Justice and the Department of Energy. After a year's investigation, each department reported there was no credible evidence that gasoline shortages resulted from collusive conduct by oil companies.

But this lack of evidence has done little to deter those revisionists of history who persist in irresponsible accusations of worldwide criminal conspiracies and collusions.

Such persistence justifies another look at both the facts and the critics themselves. The facts are on record; there is no mystery about what occurred. Here is the chronology:

• As 1977 ended, the free world's inventories of oil were at an unprecedented high—an accumulation that had been built up for a number of reasons, not least a fear that OPEC prices might be raised again at the beginning of 1978.

• The drawdown of inventories and the reduction of OPEC crude production that occurred during the first half of 1978 were neither ominous nor mysterious. Rather, they were reasonable and normal reactions to plentiful supplies.

North Sea production was increasing. Alaskan crude oil supply was increasing, too. Spot market prices were soft, and the new year 1978 had not brought the anticipated OPEC price increase. The tension that had pushed inventories so high began to ebb, with the inevitable result that OPEC production levels eased back and inventories came down. Consumers and oil companies alike felt comfortable about future supplies. Demand for gasoline rose past the forecast levels and remained high.

IRANIAN TURMOIL

• Then, in the fall of 1978, civil disturbances erupted and

the oil workers went on strike in Iran, which had been exporting between five and six million barrels of oil per day to the
free world. As turmoil increased, oil production declined
sharply, and on December 22, 1978, crude oil exports from Iran
stopped completely.

This sudden loss of some 10 percent in free world oil supply
shocked the world. Before the first disruption in oil exports
from Iran, free world inventories had been near normal levels.
But once Iranian production began toppling, it became extremely difficult to rebuild them, for the atmosphere of
impending shortages made crude oil supplies more precar-
ious and more valuable.

"MEDIA PROFITS EXCEED 'BIG OIL'"

*Over the past few years, a lot of epithets and anger
have been directed at the energy producers in the
United States...*
*At the center of all the criticism have been the profits
of energy producers...*
*In its annual study of how profits are faring in various
sectors of the U.S. economy, Forbes magazine study
found that over the past five years the energy companies
had to take second and third place to both broadcasting
and publishing — in other words, the media, the energy
companies' most vocal critics.*

From the *Birmingham News*, 1980.

Traders began to compete for oil, and prices in the spot
market started to climb steadily upward and eventually
reached $40 per barrel in late 1979. Anxiety fed on anxiety: No
one knew how long Iranian production would be lost; nor
whether other producing countries would make up the
shortage; and if they did, how long they might be willing to do
so, and at what price.

• In the first quarter of 1979, crude oil production available
to the free world fell to 50.5 million barrels a day, a drop of 1.8
million barrels a day from the levels of the fourth quarter of
1978.

THE CRISIS

Thus, the stage was set for the crisis that ensued: a world-wide production shortfall of almost two million barrels per day; a worldwide scramble for secure oil supplies that led to steadily rising prices on the spot market; US government policy that discouraged the major American oil companies from entering the spot market in search of more supply; President Carter's early initiative to establish a heating oil inventory-building program that required reductions in gasoline production during the prime driving season; and all of this complicated further by federal regulations for allocating gasoline supplies that were based on outdated demand patterns and resulted in shortages and long gasoline lines in certain areas of the country, particularly metropolitan areas.

Our critics now argue that since total free-world crude oil production was higher in first-quarter 1979 than it had been in first-quarter 1978, there really wasn't any shortage at all. This argument intentionally ignores the fact that demand had risen substantially (to 55.7 million barrels per day, the highest quarterly consumption in history), and that crude oil production had been unusually low (about 50 million barrels per day) in early 1978. It also ignores the vast accumulation of evidence demonstrating that the comparatively higher volume produced one year later was still, quite obviously, not enough.

Despite the clear record of history, the oil industry is besieged by a small cabal of critics who are determined to rewrite what happened to make it conform to their theories and not to the facts, insisting — against all the evidence — that there must have been a worldwide conspiracy among the major international oil companies to inflate the price of oil.

There is nothing puzzling about their goal, which is to break up the oil industry or bring it even further under government control. What is puzzling is the ready market they find for their discredited theories, which are published or broadcast by the communications media as though they do no great harm to the status and performance of corporations that are critically important to this nation's economic health and industrial strength.

3

DISTINGUISHING PRIMARY FROM SECONDARY SOURCES

A critical thinker must always question his or her source of knowledge. One way to critically evaluate information is to be able to distinguish between PRIMARY SOURCES (a "firsthand" or eyewitness account from personal letters, documents, or speeches, etc.) and SECONDARY SOURCES (a "secondhand" account usually based upon a "firsthand" account and possibly appearing in newspapers, encyclopedias, or other similar types of publications). A diary about the Civil War written by a Civil War veteran is an example of a primary source. A history of the Civil War written many years after the war and relying, in part, upon that diary for information is an example of a secondary source. However, it must be noted that interpretation and/or point of view also play a role when dealing with primary and secondary sources. For example, the historian writing about the Civil War not only will quote from the veteran's diary but also will interpret it. That his or her interpretation may be incorrect is certainly a possibility. Even the diary or primary source must be questioned as to interpretation and point of view. The veteran may have been a militarist who stressed the glory of warfare rather than the human suffering involved.

Instructions

Test your skill in evaluating sources by participating in the following exercise. Pretend that your teacher has asked you to write a research paper on the causes and effects of the current energy crisis. You are also asked to distinguish the primary sources you used from the secondary sources. Listed below are nine sources which may be useful in your research.

Carefully evaluate each of them. First, place a P next to those descriptions you feel would serve as primary sources. Second, rank the primary sources assigning the number (1) to the most objective and accurate primary source, number (2) to the next accurate and so on until the ranking is finished. Repeat the entire procedure, this time placing an S next to those descriptions you feel would serve as secondary sources and then ranking them. Discuss and compare your evaluation with other class members.

P or S

Rank in Importance

_____ 1. A research report by a government engineer _____
written in 1939 claiming that there is a 40
year oil reserve in the continental U.S.

_____ 2. A *New York Times* editorial attacking the _____
OPEC cartel for price gouging.

_____ 3. A speech delivered at a U.N. energy con- _____
ference by an Arab oil minister defending
the current pricing structure established by
OPEC.

_____ 4. A spectator at a U.S. Senate hearing on _____
leasing offshore lands to oil companies
describing the testimony by Michael Mc-
Closky, executive director of the Sierra Club.

_____ 5. A chapter from Albert Einstein's autobio- _____
graphy describing nuclear power as a useful
energy source.

_____ 6. A report in *Newsweek* magazine describing _____
the windfall profits earned by the oil companies when the price of domestic oil was
deregulated.

_____ 7. The article by Richard Reeves in this chap- _____
ter (Viewpoint 3).

_____ 8. A geological map followed by a 3 page _____
narrative in the Encyclopedia Britannica on
the location of known oil reserves throughout the world.

_____ 9. A history of the world energy crisis during _____
the past 10 years written by a Harvard
professor.

3

"There can be no fairness in America as long as American oil companies profit spectacularly while others suffer."

Oil Companies Exploit Consumers

Richard Reeves

Richard Reeves is a syndicated columnist whose articles appear in 150 newspapers. A graduate of Stevens Institute of Technology, he worked as an engineer for a year before beginning his career in journalism. Mr. Reeves was chief political correspondent for *The New York Times* and editor of *New York Magazine*. His books include *A Ford, Not a Lincoln* (1975) and *Convention* (1977). In the following viewpoint, Mr. Reeves states that until American consumers can be assured that they are not being exploited by the oil companies, they will not make an effort to conserve energy.

Consider the following questions while reading:
1. **What is the weapon Mr. Reeves feels the U.S. holds against Arab "oil blackmail?"**
2. **Why does Mr. Reeves claim Americans won't accept an effective conservation program?**
3. **Do you agree or disagree with the author's suggestion for an "information" commission? Explain your answer.**

Richard Reeves, "What We Need Is The Complete Truth About Oil," *The Minneapolis Star*, November 29, 1979. Copyright 1979, Universal Press Syndicate. All rights reserved.

It is stating the obvious to say that no one believes the oil companies anymore. But it seems to me worth stating that fact again because that utter and deserving lack of credibility is the core of America's energy problems—and of what is being called the crisis in political leadership.

The United States must conserve energy. If we reject a military solution—and most sane Americans do—then our major weapon against unchecked Arab oil blackmail is the fact that we are the world's largest purchaser of the liquid gold. The U.S. buys one-third of the oil produced in the world. We might have some leverage on prices if we could control our demand. Just like us, the Arabs are financially over-extended, and a reduction in American buying just might force them to face the prospect of giving up some of their pet projects, things like buying jets and building hotels, hospitals and department stores in the desert.

ENERGY CONSERVATION

But no one can seem to make us fight back. We won't really accept an effective conservation program. Why? Because our leaders are weak? I don't think that's the whole reason. I think we are, individually and collectively, unwilling to accept sacrifice unless we believe it is fairly shared and distributed.

Americans tend to revere fairness—it is, after all, preached as the underlying value of both our political and legal systems. But, a lot of us reason, there can be no fairness in America as long as American oil companies profit spectacularly while others suffer.

The companies, of course, argue that they are not profiting at the expense of the American consumer. Those 211 percent and 118 percent profit increases, they say, were made in foreign markets, are just to provide capital for exploration, really aren't as big as they seem, etc., etc.

Who believes them? Who should? What credibility does Exxon deserve? This is a company whose chairman, Clifton Garvin, Jr., says, "The benefit of the higher prices was more than offset by higher costs of both imported and domestic crude oil..."

But even superficial analysis reveals that Exxon delivers mainly Saudi Arabian oil, which it has been buying at $18 a barrel, compared to the $23.50-a-barrel oil being bought by most other oil companies.

93

The oil companies keep saying they are not making much more money than other industries. Well, look at it from the point of view of price. Domestic oil was selling for $3.50 a barrel in 1973. Now, new domestic oil is selling from $15 to $20. Either the oil companies are very inefficient and have wasted a lot of that increase, or they're making a lot of money.

Ralph Nader, consumer advocate.

94

EXXON'S LOW PRICES

Has anyone noticed how much lower Exxon's prices are than their competitors? What is happening to that $5.50-a-barrel saving? Who believes anything Garvin says? This is also a company, Exxon, that took out advertisements in both of Washington's newspapers, the Post and the Star, last June and July, saying, "In total, we will supply as much gasoline as we did last July." The facts, dug out by Rep. Herbert Harris of Virginia, were that Exxon actually cut its June deliveries by 34 percent and July deliveries by 20 percent in the metropolitan area.

Then, when gas lines choked the city, the oil companies and the Department of Energy blamed "panicky" drivers for "topping off" their tanks.

Lying? Obfuscation? Public relations? Government incompetence? It doesn't matter anymore. Americans have got to have independent oil information to forge a consensus for conservation — or for some sort of national effort to escape permanent bondage to the lords of petroleum, foreign and domestic.

INFORMATION CRISIS

Like it or not, only the government is big enough to deal with this information crisis. The president should consider appointing a respected and bipartisan commission to report to the nation on what the situation is — and that commission should have extraordinary powers. Why shouldn't "flying squads" of engineers, accountants and experts of various kinds, maybe even journalists-on-leave, with subpoena power, literally and legally invade the fields and account books of the companies? Drill exploratory holes if they have to, drain tanks, do whatever has to be done to get at the truth.

The truth will go a long way toward making us free. Maybe Exxon et al, are telling the truth. Maybe not. Maybe we are being ripped off. But we have to find out because no national program can work unless it is perceived as being fair, and we won't believe it's fair unless and until someone can convincingly report to the nation on what the hell the oil companies are really doing to us, or for us.

"The oil companies are not creating a shortage to increase prices."

Oil Companies Do Not Exploit Consumers

Richard L. Lesher

Richard L. Lesher has been president of the U.S. Chamber of Commerce since 1975. He holds degrees from the University of Pittsburgh, Pennsylvania State University and Indiana University. The author of a syndicated newspaper column, Mr. Lesher has been an educator, a consultant to NASA and president of the National Center for Resource Recovery. In the following viewpoint, he claims that the government must share the blame for sudden oil shortages and rising energy prices and insists that oil company costs eat up most of the income from increased prices.

Consider the following questions while reading:
1. **What does Mr. Lesher claim the government did that "planted the seeds" of the oil shortage?**
2. **How have government regulations reduced refining capacity, according to the author?**
3. **According to Mr. Lesher, why do oil companies only make a two cent net profit on each oil dollar?**

Richard L. Lesher, "Those Oil Profits — Who's Ripping Off Whom?," *Human Events,* May 26, 1979. Reprinted by permission from *Human Events.*

It's spring, America is in full blossom and the countryside grows more beautiful by the day. But this year, many Americans have been distracted from that beauty by an increasingly ugly reality—the energy crisis, which has given way to longer lines, higher prices and hotter tempers at the gas pumps.

The public is looking for explanations and leadership. But it's gotten little more than hot air—irresponsible rhetoric suggesting our energy crisis might disappear if enough Americans simply go after the petroleum industry with a hatchet, and if enough of them form vigilante groups to spy on each other and force down "unreasonable prices." I do not have all the answers, but let me share with you a few *facts*.

FACTS ABOUT SHORTAGE

First, the oil companies are not creating a shortage to increase prices. A real shortage already exists. The seeds of that shortage were planted in late 1971 when the federal government—supposedly to "protect" consumers—decided to artifically control domestic oil prices.

The effect was precisely the opposite because through its decision the United States government flashed the following message to the rest of the world: We don't care how much it costs to drill deep wells and explore for oil. (An off-shore installation can cost hundreds of millions of dollars.) The price our domestic producers receive will remain fixed, and we're going to rely on the good will of foreign oil producers to ensure the balance of our energy needs at reasonable prices.

Result? As costs rose, domestic exploration predictably began drying up. The United States became progressively more dependent upon the OPEC cartel, the Organization of Petroleum Exporting Countries. And OPEC has vividly demonstrated that it has the gift of grab. OPEC now jacks up prices (an increase of approximately 500 per cent per barrel in six years), and manipulates supplies almost at will. The United States is dangerously vulnerable.

U.S. VULNERABILITY

Our immediate problems pinpoint that vulnerability. The revolution in Iran led to a temporary but complete interruption in that country's oil exports. Those exports have now been resumed, but not at their prior levels. That means the United States is now receiving about 700,000 barrels a day

97

less than it normally would, which represents a shortfall of about 3.5 per cent of normal supplies. Not much, you might say, and you'd be right. But look what else is happening to aggravate the problem.

While we are seeing a *reduction* in the supply of crude oil, we are also witnessing a *tremendous increase* in the demand for gasoline—demand for unleaded gasoline is up more than 20 per cent over last year. The government has mandated the increased use of unleaded fuel, even though it gobbles up more oil per barrel in the refining process than does leaded fuel.

GOVERNMENT REGULATIONS

And to make matters worse, expensive government price and environmental regulations have sharply reduced our refining capacity. *Only one* major refinery has been built anywhere in the United States for six years, and none on the East Coast for 20 years. What a way to run a country that runs on gasoline!

HIGH COSTS HURT COMPANIES

It's a curious fact that much ado is made in the public press about the soaring costs of oil and natural gas, but virtually nothing is said about the much greater, more massive increases in the cost of extracting and processing that oil and gas. Perhaps what our industry needs is its own Nader's Raiders to get out and tell that story.

J. Hugh Liedtke, Chairman of Pennzoil Co., *Vital Speeches of the Day*, June 15, 1981.

The federal government also insists refineries process a greater share of their crude oil into home heating oil than into gasoline stocks—just the opposite of what usually happens at this time of year. Hopefully this additional sacrifice of gasoline supplies will at least prevent schools, offices and homes from being without heat.

Finally, the oil companies have had to voluntarily reduce

THE OTHER OPEC

the amount of gas they deliver to their stations under contract to try to lessen the impact of the gasoline shortage. Stations are being allocated gas on the basis of the quantity of gas they sold during 1977–78, not on the basis of their overhead. That means self–service stations might receive more gas than full–service stations, even though the latter need to sell more gas to cover their high overhead and stay in business. These are just some of the reasons why lines at service stations are lengthening and prices are rising...

Who *really* makes the money on the gas you buy? From the time the oil leaves the sand in the Mideast to the moment it fills up your tank as gasoline, all governments, both foreign and domestic, can take up to 65 cents of every dollar you pay. And after subtracting costs for refining, storage, transportation, marketing and local dealers, the oil companies are left with about two cents net profit. The real question Americans should be asking is: Who's ripping off whom?

UNDERSTANDING STEREOTYPES

A stereotype is an oversimplified or exaggerated description of people or things. Stereotyping can be favorable. However, most stereotyping tends to be highly uncomplimentary and, at times, degrading.

Stereotyping grows out of our prejudices. When we stereotype someone, we are prejudging him or her. Consider the following example: Mr. X is convinced that all Mexicans are lazy, sloppy and careless people. The Diaz family, a family of Mexicans, happen to be his next–door neighbors. One evening, upon returning home from work, Mr. X notices that the garbage pails in the Diaz driveway are overturned and that the rubbish is scattered throughout the driveway. He immediately says to himself: "Isn't that just like those lazy, sloppy and careless Mexicans?" The possibility that a group of neighborhood vandals or a pack of stray dogs may be responsible for the mess never enters his mind. Why not? Simply because he has prejudged all Mexicans and will keep his stereotype consistent with his prejudice. The famous (or infamous) Archie Bunker of television fame is a classic example of our Mr. X.

Instructions

Read through the following list carefully. Mark S for any statement that is an example of stereotyping. Mark N for any statement that is not an example of stereotyping. Mark U if you are undecided about any statement. Then discuss and compare your decisions with other class members.

S = Stereotype
N = Not a stereotype
U = Undecided

_____ 1. Most scientists agree that nuclear power can solve the energy crisis.

_____ 2. Most scientists disagree on which energy sources should be developed for future use.

_____ 3. As a rule, college students are opposed to the use of nuclear power plants.

_____ 4. Poor people are not hurt by higher gasoline prices because they don't own cars.

_____ 5. Economically, rising energy prices probably hurt poor people more than wealthy people.

_____ 6. Sometimes energy producing companies make an effort to restore or improve the environment disturbed by energy exploration.

_____ 7. Oil company executives are merely interested in profits, not the public good.

_____ 8. Profits, not good will, are the chief concern of oil companies when they explore for oil in Third World countries.

_____ 9. Most member states of the OPEC cartel will generally price oil at whatever the "traffic will bear."

_____ 10. Many consumers despise OPEC and the oil companies for their exorbitant price increases.

_____ 11. Utility companies always request higher rate increases than are actually warranted.

BIBLIOGRAPHY

The following list of periodical articles deals with the subject matter of this chapter.

Richard Corrigan — *The Era of Cheap Energy Rides into the Sunset,* **USA Today**, May, 1980, p. 11.

C.E. Curtis — *Natural Gas Pains,* **Forbes,** June 23, 1980, p. 61.

J. Egan — *Crude Politics: What America Doesn't Know About Oil,* **New York**, March 19, 1979, p. 15.

M. Stanton Evans — *"Allocations" Program Causes Gas Lines,* **Human Events,** July 7, 1979, p. 10.

Pamela Haines and William Moyer — *"No Nukes" is Not Enough: The Need for a New Energy Strategy,* **The Progressive,** March, 1981, p. 34.

Chris Hedges and Carlos Moralis Troncoso — *What is Gulf and Western to the Dominican Republic?,* (opposing viewpoints), **The Christian Science Monitor**, January 17, 1979, p. 22.

Robert J. Henle — *Energy Policy: A Matter of Life or Death,* **America**, February 14, 1981, p. 121.

G.H. Lawrence interviewed by T.A. Cohen — *Natural Gas,* **Forbes,** January 5, 1981, p. 209.

R.G. Lugar — *Energy: The Fallacy of Controlled Scarcity,* **Vital Speeches of the Day,** June 15, 1977, p. 520.

Nation — *Big Oil or Us,* January 5, 1980, p. 3.

William E. Simon — *Tilting at Windfall Profits,* **Human Events**, December 1, 1979, p. 14.

Time — *After the Chill Comes the Bitter Bill,* February 28, 1977, p. 46.

Chapter

What Are
Our Energy
Alternatives?

"No matter how high the price of oil has been, synfuel always has cost more."

Synfuels Are Too Costly

Morton Kondracke

Morton Kondracke has been a reporter for the *Chicago Sun Times* since 1963 and a member of its Washington bureau since 1968. In the following viewpoint, he claims that the commitment to synfuels will create environmental and economic havoc and suggests alternatives such as conservation and solar energy.

Consider the following questions while reading:
1. **Who does Mr. Kondracke claim will profit from synfuel development?**
2. **What does the author feel is the key problem with developing synfuels?**
3. **What alternative to synfuel does the author advocate?**

Morton Kondracke, "Synfuel Madness," *The New Republic,* July 21-28, 1979.
Reprinted by permission of *The New Republic,* © 1979 The New Republic, Inc.

105

Gas lines, the OPEC price increase, and the troubles in Iran and at Three Mile Island finally have convinced Congress that there is an energy crisis, and that's good. But Congress's response to the crisis — or rather, its response to the fear that the public will punish it for not doing something sooner — has been to latch on to an ill-conceived crash program to develop synthetic fuels... For no well-considered reason, the leaders of the US government seem about to commit the country to a huge new enterprise that will stir men's souls and divert their attention, but is likely to be an economic, social, and environmental disaster. The synfuel crusade will put vast sums of money into the hands of US energy companies and other vested interests. It will require government or consumer subsidies, probably forever. It will raise interest rates and otherwise fuel inflation, create environmental havoc, and shift resources from less macho alternatives — active conservation, worldwide exploration, and solar energy — which show more promise...

SYNFUEL BOOSTERS

Behind the scenes, powerful interests are boosting synfuel. The oil companies might be thought hostile to potentially competitive artificial fuel, but in fact they own 20 percent of the country's coal. The three main US techniques for liquifying coal into oil are being developed by Exxon, Gulf, and Ashland Oil Company. Mobil is into making methyl alcohol out of coal, and two of the big oil shale developers are Union Oil Company and Occidental. Besides energy companies, the auto industry loves synfuel, which might relieve the pressure for mileage standards... The United Auto Workers presumably will support synfuel if this means its members can keep on making big cars. The United Mine Workers foresees a huge increase in membership from synfuels, and the AFL-CIO construction trades unions anticipate a boom in employment building synfuel plants. If the federal government is going to guarantee loans of $100 billion, $200 billion, or $300 billion, banks and investment houses won't complain, nor will big law firms, which will write all the contracts...

Unfortunately, there are problems. The key one is economic. Hitler's Germany produced gasoline from coal during World War II, and South Africa plans to get 45 percent of its energy supplies from synfuel (using, of course, cheap black labor in the coal mines). But no other country in the world besides those two has chosen to go the synfuel route because, no matter how high the price of oil has been, synfuel always has cost more... It's no sinister accident that synfuel

106

costs stay ahead; it takes energy to convert coal into oil or gas, and energy costs money. As a result, private companies have seen no profit in synfuel production, and the only plants existing in the US at the moment are small experimental operations.

FEDERAL SUBSIDIES

If the United States decides on a big synfuel push, the federal government not only will have to help companies build the plants, but also will have to provide permanent subsidies to induce people to use synfuel instead of oil. The subsidy could be honest and direct, and show up in the federal budget. More likely, the government will require the companies to buy the stuff, and let them pass the cost on to consumers. How much all this will cost the US economy is not clear, but it will be billions...

SYNFUELS EXPENSE TOO GREAT

Overall, the astronomical expense, environmental degradation, and long lead time involved in synfuels production set synthetics apart from petroleum — except that they too are based on fossil fuels, so the problems associated with their use are all too "conventional." True, much lower-grade and abundant ores can be used in the manufacture of synfuels than can be used in conventional fuel production but the expense of extraction far outweighs the bargain of the fuel stocks. The ultimate illogic of synfuels is that to preserve an oil-based economy whose days are clearly numbered, we would have to invest more of the capital needed to retool for another era.

Kathleen Courrier, *engage/social action*, May, 1980.

Some synfuel advocates say that all the risks and costs are worth taking in order to be free of OPEC. But there are alternatives. If the US goal is to replace five million of the nine million barrels of oil we import per day, we can do it cheaper and more safely. One way is by active conservation — or "energy productivity," as some PR-minded advocates call it. These advocates claim that by insulating homes and replacing insufficient heating units, the US could save two million barrels a day by 1990 and four million by 2000. Additional savings of three million barrels a day can be achieved in the industrial and commercial sectors. These conservation measures do not involve reducing US growth or industrial production. To the contrary, refitting industries and weatherizing homes would provide new jobs and increase the gross national product. It is not a no-growth, un-American idea at all.

NON-OPEC OIL

Another alternative to OPEC oil is non-OPEC oil, including Canada's heavy oil. Mexico, China, and other non-OPEC producers probably will charge the OPEC price for their products, but it's better to pay them than to pay OPEC. The more oil there is available in the market, the less OPEC can hike prices. As long as Congress is in a mood to spend money on energy, it should establish a fund to finance exploration abroad by independent US oil companies. The US should also try to write contracts with non-Arab oil producers guaranteeing the US a supply at world prices, regardless what happens in the Persian Gulf.

The federal government also should put more money into solar energy, including conversion of organic biomass into fuel. The Harvard Business School's energy project concluded that the equivalent of four million barrels a day of oil could be produced from solar sources by 1990.

Congress...should proceed with tests of synfuels to see whether they can be made commercially sound. But to commit the country to a massive crash effort is irresponsible. The government would be selling the country a pig in a pork barrel if it did so. Done out of political panic, it is the government equivalent of topping off your tank.

"Fuels made from shale and oil sands and IBG from coal...are estimated to be producible at costs comparable to the current cost of imported oil."

Synfuels Can Be Affordable

The Lamp

The Lamp is a magazine published quarterly by the Exxon corporation for its shareholders. Produced by Exxon's Public Affairs Department, Communications Programs Division, the magazine is available to anyone upon request. In the following viewpoint, *The Lamp* claims that synfuels are a viable alternative to imported oil and, while recognizing that production problems do exist, concludes that the day of synfuels has arrived.

Consider the following questions while reading:
1. **According to *The Lamp*, what are the reasons that synthetic fuels have been in demand for two centuries?**
2. **What are some of the reasons offered to explain why synthetic fuels are a better alternative than imported oil?**
3. **What types of cooperation does *The Lamp* believe will be necessary to build a synfuels industry?**

"Synthetic Fuels: The Processes, Problems and Potential," *The Lamp*, Summer, 1980. Reprinted by permission from Exxon Corporation © 1980.

Synthetic fuels—that is, liquid and gaseous fuels produced from coal, oil shale, oil sands or plant matter—have been providing energy bridges and stop-gaps for nearly two centuries. Sometimes demand for synthetics has been based on cost advantages over other energy forms, and sometimes because no practical alternatives were available to fill a need...

SYNFUELS IN THE SPOTLIGHT

Today, synthetic fuels are back in the spotlight as energy consumers look for ways out of a worldwide energy crisis. While there are currently only a handful of commercial plants, notably in Canada and South Africa, the economics of synfuels are improving rapidly as the cost of imported oil spirals higher. "It is increasingly clear that large quantities of synthetic fuels will be needed in the decades ahead," says Manny Peralta, manager of the coal and synthetic fuels division of Exxon. "They can provide an essential bridge to help the world make the transition from primary dependence on fossil fuels to that day when we can draw most of our energy from nondepleting sources."

Oil and gas provide almost 75 percent of total energy supply in the United States, and a comparable percentage in the rest of the industrialized world...

Even with the expected continued improvement in energy conservation, the arithmetic of population growth and economic expansion indicates that by the year 2000 the world will need about two thirds more energy in absolute terms than it consumes today. Most of this growth in energy demand will have to be covered by the direct burning of coal and by the nuclear fission of uranium. Renewable or non-depleting sources such as hydroelectric power or solar energy will be able to contribute only a small fraction of energy supply...

To meet this demand entirely from imported petroleum would require doubling the present level of U.S. oil imports. The economic and political price of this might well be more than the nation would want or could afford to pay—even assuming that exporting countries would make the oil available.

SYNTHETICS: A BETTER ALTERNATIVE

Synthetics can offer a better alternative. Some forms are already competitive, and as world oil prices rise still higher (which most experts believe is inevitable), synthetics are expected to come into their own as commercial fuels. The distribution systems for gas and liquid fuels are already in place, and the raw materials are available in abundance. Coal and oil shale in the United States, even after deducting the coal to be used for conventional purposes, could yield synfuels with the energy equivalent of a trillion barrels of oil—enough to sustain a 15-million-barrel-a-day production rate for 175 years.

'Help. . .if I don't find oil pretty soon I'll freeze'

LePelley in *The Christian Science Monitor* © 1980 TCSPS.

The United States has the capacity to build a synthetics industry that could supply a fourth of the nation's demand for liquid and gas fuels—the equivalent of 6 million barrels of oil a day—by the end of the century, and perhaps 15 million barrels a day by the year 2010...

Coal-based liquids and SNG are still somewhat too costly to be competitive with imported petroleum, though they will probably cross that economic threshold before long. Fuels made from shale and oil sands and IBG from coal, on the other hand, are estimated to be producible at costs comparable to the current cost of imported oil.

Whether a synfuel plant is processing coal, oil shale or oil sands, economics dictates that it be very large; a single commercial plant should produce the equivalent of around 50,000 barrels or more a day of liquid fuels. A 50,000-barrel-a-day oil shale mine and plant or coal liquids plant starting up in the late 1980s will cost over $3 billion.

The cost of building an entire synfuels industry from scratch will run into hundreds of billions of dollars in the United States alone. But companies like Exxon have demonstrated their belief that the job is manageable and that the investments are worth making. Exxon has already spent or committed close to $1 billion worldwide for synfuels research, resources and construction. Synfuel projects now under active consideration could add as much as $15 billion to company investments in the decade ahead.

One reason for high costs is the amount of time needed to convert scientific theory into reality in the synfuels field, since every day of delay in starting a project adds about $500,000 to the ultimate cost due to inflation...

COOPERATION TO DEVELOP SYNFUELS

Building a synfuels industry the right way will call for an exceptional degree of cooperation among industry, government and private citizens. In the United States, government support of synfuels research and development has already speeded the advent of commercially acceptable processes. A critical need now is timely resolution of many public policy questions that cast uncertainty over major development efforts. The needed transfer of water, for example, cannot occur without government involvement and broad-based public support. Most coal and shale resources are on federally owned land, and decisions are needed to accelerate the leasing process. The nation needs a regulatory climate which will encourage development on a reasonably rapid basis, making allowance for the long lead times inherent in synfuels development. A recent study by the Department of Energy found that existing regulations governing air and water quality would limit the potential size of a U.S. synfuels industry to something on the order of two million barrels a day. If we want synfuels to make a larger contribution, we will have to accept some modification of environmental standards or improve our technology for the control of pollutants. Most likely, we will have to do both...

SYNFUELS DEVELOPMENT IS CRUCIAL

A coal-based synfuels program has enormous potential. It would use both established and new technologies to convert our coal reserves into a variety of environmentally acceptable fuels. Processes of this general type have been successfully used elsewhere in the world for the past 40 years...

Swift development of synthetic fuels is crucial to the United States. It would be a strong signal to OPEC — and a welcome one to our allies — that America is commited to the goal of eventual energy independence. As synfuels come into the market and cut our need for foreign oil, they will impose a price ceiling on energy, a ceiling that will be determined by market forces instead of the cartel's political decisions.

Edward Donley, *Vital Speeches of the Day,* June 1, 1980.

The day of cheap and plentiful petroleum is clearly over. Just as clearly, the day of synfuels has arrived.

"The transition to solar energy is essential to sustain employment as a whole."

Solar Energy: Hope of the Future

Barry Commoner

Barry Commoner is a professor of Earth and Environmental Sciences and Director of the Center for the Biology of Natural Systems at Queens College, Flushing, New York. A PhD from Harvard University and an LLD from Clark University, he was the 1980 presidential candidate for the Citizen's Party. Dr. Commoner is Chairman of the executive committee of the Scientists Institute of Public Information and has authored several books including *The Politics of Energy* (1979). In the following viewpoint, he responds to a series of questions regarding his published view that the U.S. must develop its energy from renewable resources, particularly solar. The interview was conducted by Roy J. Rotheim, Executive Editor of *Challenge* magazine.

Consider the following questions while reading:
1. **What energy producing methods does Dr. Commoner include when he speaks of "solar energy?"**
2. **Why does the author claim that solar heating is already cheaper than conventional heating?**
3. **What does Dr. Commoner believe is the danger of the present nonrenewable energy system?**

Barry Commoner, "The Case for Solar Energy." Reprinted by permission of *Challenge* The Magazine of Economic Affairs, September/October, 1979. Published by M.E. Sharpe, Inc., Armonk, NY.

Q. In your recently published book, *The Politics of Energy,* you make a strong case for the United States to develop its energy from renewable sources—particularly solar—rather than relying on nonrenewable sources such as oil, natural gas, coal, and uranium. What is meant by solar energy?

A. Solar energy is any energy that comes from the sun. It includes heat directly collected from sunshine striking a suitable absorber. That's usually called a solar collector. It also includes wind, because wind is caused by the rising of solar-heated air and cold air rushing in underneath. It includes hydroelectric power because the sun raises the water to the mountaintops from whence it falls. It includes all organic matter produced concurrently by plants now growing. That means wood, all food and fuel derived from crops, or alcohol derived from grain. It also includes everything derived from food such as sewage and manure. And there is also the direct conversion of sunlight into electricity, for example, by photovoltaic cells.

Q. President Carter, in his energy address on July 15, proposed the creation of a solar bank so that by the year 2000 we can derive 20 percent of our energy from solar power. Is this goal realistic?

A. It certainly is. In fact, by the year 2000 it could easily be surpassed. It happens that I proposed the solar bank for the first time two-and-a-half years ago in a speech to the National League of Cities. The idea was to set up a bank in each city that would give favorable loans for solar equipment, particularly solar collectors for householders. My initial concept was that it would take about 50 years to make the full transition to solar power.

Q. How would we make that transition?

A. In the first place, we must recognize that there are some fuels for which we have no alternatives in the foreseeable future—liquid fuels, for example. Ultimately, solar liquid fuel, such as alcohol produced by way of biomass conversion from grains, will be required to drive perhaps one-third of the land vehicles; the rest would be electrified. Meanwhile, we still require some bridging fuels. Natural gas would be the best. Natural gas-fired cogenerators, in which the fuel is used to produce both useful heat and electricity, would be installed wherever possible, coupled with more extensive gas distribution systems built to supply them. Also, present installations that use electricity for heating would be replaced by gas-fired cogenerators. During the first 25 years of transition, the production of solar energy would increase with installation of

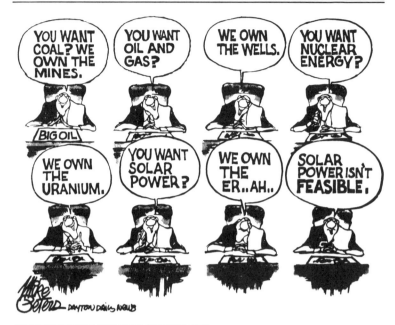

Reprinted by permission of Mike Peters, *Dayton Daily News*.

solar collectors as well as the use of photovoltaic cells and wind generators. By the twenty–fifth year, if we had effective conservation, production of solar energy would amount to perhaps 50 percent of the total U.S. energy budget. During this time, it would be possible to eliminate all oil imports without increasing the present rate of producing domestic oil, coal, and nuclear power.

Q. What would happen during the second 25 years?

A. Then our goal would be to have 90 percent solar energy and 10 percent natural gas. Within this period, it would be necessary to make sufficient gas available. But that is clearly possible.

I should add that since my book was published there's been some evidence that we can speed up the transition to solar energy.

Q. How?

A. Well, for example, at the Center for the Biology of Natural

Systems at Washington University, our staff has just finished a survey of the potential for introducing alcohol from midwestern agriculture. This involves the changing of crop patterns so that there's excess carbon to make into alcohol. By the year 2000 we could be producing, without introducing any new fermentation technology, 50 billion gallons of alcohol a year, which is about half the present gasoline consumption. By simply looking into the capability of a particular solar procedure like that, I think that in 20 to 25 years we could be half solar.

Q. But is such a process economical?

A. Yes, if you can find the front-end investment, it's bound to be. My basic argument is that as long as we continue to use nonrenewable fuels, the cost of energy will rise exponentially. That is what you economists call a maintenance cost; it is a drain on the output of the economic system, which is simply devoted to keeping it going. What we have is a situation in which the economic system is being cannibalized by the exponential increase in the cost of this one input.

Q. What is your evidence that the cost of nonrenewable fuels will increase exponentially? And why isn't the same thing true for renewable fuels?

A. Look, it all boils down to this: as a nonrenewable energy source is depleted, the cost of producing it will rise faster and faster with each added unit of energy produced. In 1972, a National Petroleum Council study, which by the way has been confirmed in more recent work, showed exactly that. The study assumed that between 1970 and 1985, the annual increase in the production of domestic oil would be 18 percent. Under this assumption, the selling price of the oil (in 1970 dollars), based on expected production costs with a fixed rate of profit, would need to rise from $3.18 a barrel in 1970 to $3.70 in 1975, $5.16 in 1980, and finally to $7.21 by 1985. If these conclusions were extrapolated further, you'd get $14 a barrel in 1995, $70 in 2020, $380 in 2045, and $2,000 in 2070.

The investment of capital in the production of energy is moving in the direction of more and more capital per unit of energy yielded. Between 1972 and 1974, the energy industry employed 15 percent of the capital available for all business investment. But projections for 1975-82 visualize the energy industry increasing its share to 25 percent.

This isn't the case, however, with the implementation of renewable energy. In the cases of solar space heat and hot water we already know from very direct economic computations that for the central part of the country it is economical today to make the investment based on a 15-year amortization of standard loan rates, given the exponential rise in cost of the conventional alternative form of energy.

Compare two kinds of heating systems—a conventional system and a solar system. The conventional system requires

Barry Commoner

a relatively small initial cost, about $2,000 to $4,000 for a furnace, maybe a little less for electric resistance heating. However, the cost of fueling this investment is increasing at an exponential rate. Compare this to the installation of a solar system. Here the initial cost is higher, roughly $15,000 to $20,000 for a home in the central part of the country. But in this case, the fuel, that is the sun, is free. Except for standard maintenance costs, there is no cost for heat. So comparing the two systems over a 15-year period, the solar is already cheaper...

Q. A recent study on the "Employment Impact of the Solar Transition," prepared for the Joint Economic Committee, takes the position that there are fewer jobs in a big power plant than there would be in building solar collectors. Do you agree?

A. I think that's right. But I would say that the employment impact of the solar transition is a much more general thing; that is, unless you can sustain the economy and allow it to grow, there will be no jobs in either sector. As far as I'm concerned, the danger of the present nonrenewable energy system is that I don't see any way in which the economy can survive it. Therefore, the transition to solar energy is essential to sustain employment as a whole. And I think that's the way I would argue in terms of energy and in terms of jobs. When I have talked with union people, some of them have taken this attitude: Look, you need energy to run the factory, you need the factory to have jobs. Therefore, anything that cuts down on the availability of energy generally is bad. That's why we're against the closing down of nuclear power plants. The argument that I make is that not all energy is the same with respect to its impact on employment and environment. Union people have to look at alternative ways of producing and using energy and ask themselves how it affects their interests. Notice I didn't say that solar collectors would create more jobs than nuclear power would. Rather, I took the more general approach and said that capital invested in nuclear power is capital wasted, so to speak. In that sense the economic yield per dollar invested is low because of the high capital cost of nuclear power. What this does is prevent the entrepreneurs from having capital available to build factories and create jobs.

119

"Proposals for a major conversion to direct solar energy are at best proposals to reverse thousands of years of development."

Solar Energy: A False Hope

Samuel McCracken

Samuel McCracken is assistant to the president of Boston University. He is the author of *The War Against the Atom*, scheduled to be published by Basic Books in 1981. In the following viewpoint, Mr. McCracken attacks Barry Commoner's views on solar energy and raises questions as to the feasibility of such an energy source.

Consider the following questions while reading:
1. According to the author, Barry Commoner's "solar scheme" depends upon what four major developments?
2. According to Mr. McCracken, how does solar power generate wastes?
3. According to the author, by what method does Commoner plan to dispose of the present energy supply companies?

Barry Commoner has long been a solar advocate — *The Poverty of Power* was largely a eulogy of the sun's potential — and his new book, *The Politics of Energy*, is already very influential. It makes about as strong and detailed a case for a "solar" economy as can be imagined. Yet it is fatally flawed, not only in its technical argument for solar energy, but in the ideological skeleton that lies just below the technological skin.

COMMONER'S SOLAR SCHEME

Commoner begins with three chapters that are really tangential to his theme; their function is the prophetic one of denouncing the feasts of the Samaritans vehemently enough to convince us that his own very expensive and constraining proposals are an improvement. The chapters take the form of a critique of the present national energy policy, an exercise roughly equivalent to machine-gunning drugged fish in a barrel. Having demonstrated once again that we cannot rely indefinitely on nonrenewable resources, he sets up the alternative possibilities of an energy system based on the breeder reactor as against one based entirely on solar energy in its widest sense. Since neither system could be set up immediately, he proposes a "bridging fuel" for each. For the breeder system, this bridge is coal, and for the solar system, natural gas.

Not surprisingly, Commoner rejects the breeder-based system, primarily on the conventional grounds of safety and the risks of proliferation... To the conventional ideas he adds the objection that a breeder-based economy would require a major expansion of the use of coal, which he concedes to be a serious environmental, occupational, and public-health threat.

The solar scheme rests on four major developments. The first of these is the rapid deployment of solar technology for water and space heating. Commoner believes that this technology has already arrived, is now competitive, and need merely be put to work, normally in conjunction with existing systems. Additionally, there would be a rapid development of photovoltaic technology to allow the conversion of sunlight to electricity through individual installation on houses. This technology Commoner believes to be on the verge of success, requiring only mass purchases by the government in order to become economically competitive with central-station electricity. Next, we would build a new alcohol industry capable of producing some 50 billion gallons of liquid fuel a year, displacing about half of our present gasoline supply. Finally, there would be a vast new methane in-

Reprinted by permission of the Chicago Tribune – New York News Syndicate, Inc.

dustry, producing that gas from various feedstocks, and shipping it around the country through a major expansion of the present natural gas network...

PROBLEMS OF SOLAR ENERGY

Contrary to popular belief, direct solar power generates waste, not in the production of power, but in the production of its materials. Thus, as Petr Beckmann has noted, the production of the steel and glass and aluminum in solar plants uses substantial amounts of energy, almost all of it derived from fossil fuels, and all of it generating wastes, some of them radioactive. Moreover, the industrial processes involved here would generate their own wastes. Commoner, along with many other direct-solar enthusiasts, decries large central solar plants; if 1000 megawatts were to be generated by many small residential units, the waste, Beckmann points out, would be even greater.

Solar power also raises serious problems of efficiency. Photovoltaic generators are, from the point of view of capital investment, much the least efficient means to generate electricity. They are not available after sundown, and so have an "availability factor"—the proportion of time a generator can

122

work—much lower than any thermal plant...

But the technical problems of solar energy are substantially less interesting than its social ones. Lying just below the surface in most discussions of alternative energy are proposals for reshaping how we live by controlling the type and amount of energy available to us, as well as the way it is supplied to us. These are especially notable with direct solar energy.

There are serious constraints implicit in decentralized solar energy. One of these affects architecture: at minimum, each house would have to be oriented with its major roof area facing south. It would not matter what is dictated by the lay of the land, because these considerations would be subordinated to the simple problem of getting energy. And so, ultimately, with construction on neighboring lots: the sun comes into North America at a fairly low angle, and if the right of one householder to an all-solar home were to be preserved, his neighbors to the south would not be able to build very high next to him, and indeed their access to trees might be sharply limited whether they wished to go solar or not...

In short, proposals for a major conversion to direct solar energy are at best proposals to reverse thousands of years of development by which man, through the division of labor, has made the acquisition of energy increasingly the province of fewer and fewer increasingly well-paid specialists. This is what the Commoners are decrying when they call for the decentralization of energy sources...

A NAIVE ASSESSMENT

He ends his book by surveying the political problems inhering in the grand transition. One could summarize his argument by saying that solar energy will be good for the Peepul and less good for the bosses, except those who run energy-intensive businesses. The argument takes leave of technology and shows quite clearly the trend and quality of Commoner's ideology.

He begins with one of those breathtaking examples of naivete and self-contradiction that freight his work. We have been told (he tells us) that we are running out of oil and natural gas. No such thing, he says, for Mexico's newly discovered reserves exceed those of Saudi Arabia and we have immense amounts of "unconventional" natural gas.

Well, let us assume that the natural gas could be recovered

at economic rates. Oil, whatever its source, is just the sort of fuel that Commoner has told us we can no longer afford. There is no reason why Mexico should sell oil to the United States except at the world price and no prospect that it will. To propose that we rely on Mexican oil is to propose something Commoner elsewhere treats as intolerable, and, back in the real world, to guarantee an exacerbation of the sort of problems we now have with OPEC. If Mexican oil is more useful to us than Saudi oil, it can only be on assumptions that are sometimes called "imperialist."

THERE GOES THE SUN

Arriving free of charge, leaving no residue and making no smoke, solar energy nonetheless may turn out to be more hazardous than nuclear, more polluting than coal, and more costly to the consuming public than petroleum...

Here is Fact One: sunlight reaches the ground at a global average of 160 watts per square meter. The most optimistic engineers say that by the time we allow for variations in cloud and atmospheric cover, the natural resistance of materials, and the need to convert the electricity to alternating current, we are not likely to exceed a recovery efficiency of 5 to 10 per cent.

That computes to an average power output of about 25 megawatts per square mile. Thus the entire estimated U.S. power requirement for the year 2000 could be met by covering an area equivalent to that of the state of Oregon with solar collectors.

Donald C. Winston, *Newsweek*, December 3, 1979.

Commoner's main concern is how to dispose of the present energy-supply companies once we have erected their replacements. For the electric companies the solution is simple: competition from photovoltaic electricity would drive them into the ground financially, and we would nationalize them in order to pick up the still-useful distribution systems they control. There is no estimate of the total cost—to taxpayers and shareholders—who together would have to pay it.

"SOCIAL GOVERNANCE"

The oil companies, for unspecified reasons, he is unwilling to nationalize. They would be offered the opportunity to become public utilities like the natural-gas pipeline people. As for the coal companies, they would suffer the cruellest fate of all, for Commoner simply ignores them. Presumably they would be bankrupted without even the beneits of nationalization. And their workers? Let them distill alcohol in Iowa.

Most of the problems inherent in the grand transition Commoner would solve by something called "social governance." It is not clear what this is a euphemism for, although Commoner does reassure us that, having survived Joe McCarthy and Watergate, we can probably survive it too. There is nothing vague, however, about another key phrase. In his last sentence Commoner calls for us to adopt as a major national goal something he terms "economic democracy." Although he does not define this term here, he did so in his earlier book, *The Poverty of Power,* where "economic democracy" is explicitly and quaintly defined as the economic system of such states as China, Cuba, and the USSR—places, as is well known, in which all economic decisions are made by individuals and popularly elected assemblies...

The energy crusade as preached by a Commoner and followed by many thousands of nostalgic activists is little more than a continuation of the political wars of a decade ago by other means. Having despaired of convincing Americans that their society is the most oppressive and brutal in the world, they now tell Americans that it is the most dangerous in the world. Where salvation was once to be gotten from the Revolution, now it will come from everyone's best friend, that great and simplistic cure of all energy ills, the sun. Where it was once fashionable to argue that we had to smash the state before we could decide what to erect in its place, now the Commoners feel obliged to sketch out a rosy "scientific account" of what will ensue after the destruction of the older order. This progression is not really progress: in reality the very machinery of the state is being skillfully utilized to destroy the existing energy order long before its replacement can be prepared.

Of the late 5th-century Romans, it could at least be said that whatever their other faults, they did not *mean* to be followed by the Dark Ages.

"Solar ponds...are proving themselves to be one of the most exciting developments in the nearly desperate search for alternative energy sources."

The Solar Ponds Alternative

Michael Dorgan

Michael Dorgan is head of the California News Bureau, a news and feature service for 10 major daily newspapers. A graduate of the University of Wisconsin, Madison, he worked for the *Madison Capital Times* and was a freelance writer before accepting his current position. In the following viewpoint, Mr. Dorgan briefly explains the technology of solar salt ponds and examines the possible problems and limits to its widespread use.

Consider the following questions while reading:
1. **What types of energy does Mr. Dorgan claim solar ponds can produce?**
2. **According to the author, what possible environmental problem exists with solar ponds?**
3. **What technical problem does Mr. Dorgan claim has hindered small pond construction?**

Michael Dorgan, "Solar Salt Ponds Generate Energy Optimism," *St. Paul Pioneer Press,* December 14, 1980. Reprinted by permission of the author.

At first glance it looks suspiciously simple. Sunlight heats a pond, the heated pond water converts a liquid into a vapor, the vapor turns a turbine and—presto!—on go the lights.

At second glance it still looks simple, but no longer suspiciously so. Solar ponds—or, more correctly, nonconvective salinity-gradient solar ponds—are proving themselves to be one of the most exciting developments in the nearly desperate search for alternative energy sources.

THERMAL ENERGY

Already a half-dozen small solar ponds around the country are providing thermal energy, and now plans are being drawn for a solar pond project in Southern California that would generate 600 megawatts of electricity, enough for a city of half a million.

And that is just the beginning. No estimates are yet available from a U.S. Department of Energy study of the national potential for solar pond energy, but solar expert John Becker estimates that the potential for California alone may be as high as 33,000 megawatts. That is the equivalent of 570 million barrels of oil and represents more than 75 percent of California's present generating capacity of 40,000 megawatts.

Becker is head of research and development of thermal systems at the National Aeronautics and Space Administration's Jet Propulsion Laboratory (JPL) in Pasadena. JPL, under contract with the Department of Energy, is one of three funding sponsors for the planning phase of the Southern California Salton Sea project, the others being the State of California and Southern California Edison Co.

"Of all the new (alternative energy) technology, this is probably the most exciting," says Becker. "There's a lot of potential on a large scale."

VERSATILITY

One thing that gives solar ponds such enormous potential is their versatility. They can produce both thermal and electrical energy, and apparently can do so on almost any scale. A section of a large body of salt water can be diked off into a pond to generate multi-megawatts of electricity, or a little barnyard pond can be dug to heat farm buildings and to dry crops.

Yet versatility is not their only attractive feature. Solar

ponds are relatively simple in design and construction, they require no fuel other than sunlight and they produce no wastes.

The only environmental damage a solar pond can inflict is to leak salt water into adjacent lands or waterways. That would hardly matter in the case of large ponds constructed on bodies of salt water, like the Salton Sea, but it could be a problem with small ponds built in fertile fields or over water tables...

HOW IT WORKS

Salt is used in solar ponds to trap heat. Normally when sunlight heats a pond the heat is quickly lost through natural convection — the heated water expands, rises to the surface and releases the heat into the atmosphere.

THE POTENTIAL OF SOLAR PONDS

One thing that gives solar ponds such enormous potential is their versatility. They can produce both thermal and electrical energy, and apparently can do so on almost any scale.

But when the lower layer of a pond is heavily salted, the heat is trapped because even when expanded the highly saline layer remains denser than the overlying layer. Temperatures in the lower layer of solar ponds reach more than 200 degrees F. and, insulated by the upper layer, have been shown to retain their heat for as long as 130 days without sunshine.

If thermal heat is desired, hot brine from the bottom of a pond can simply be pumped through pipes into buildings, where it can be used for either heating or cooling. In Miamisburg, Ohio, for example, a pond is being used to heat a municipal recreational building and swimming pool.

If electricity is desired, the hot brine is pumped to a generator, where its heat converts a liquid, usually freon, into a gas. The gas drives a turbine, which generates electricity.

Once through the turbine, the gas is converted back into a liquid, using cooler water from the upper layer of the pond...

HIGH COST NOW

According to Becker, solar pond generating plants can now be built at an installed cost per kilowatt of $1,600 to $2,000. Though that cost is higher than some conventional power plants, Becker claims that when all factors are taken into account—including use rate (the time the plant is actually producing electricity), maintenance and rising fuel costs—solar plants will prove cheaper than coal, oil or nuclear-fired plants.

"When you take all costs compared to efficiency, you get (with solar ponds) one of the two cheapest sources of energy," says Becker. "Wind is the only thing that comes close."

So why haven't solar ponds been sharing the headlines with soaring fuel prices and nuclear worries?

According to Doug Elliott, head of the solar energy division of the San Francisco field office of the Department of Energy, the solar pond is such a "low-key, backyard approach that it sort of snuck up on us."

"We tend to think of solving problems by high technology rather than with something so simple," said Elliott. "One of the attractive features (of solar ponds) is that we don't see any technical obstacles. Technically everything looks feasible and straightforward."

TECHNICAL PROBLEM

One technical problem that has hindered small pond construction, says Becker, is development of a strong, durable liner. But that is a need he expects to soon be met.

Once the small technical wrinkles have been ironed out, it would seem that the only restrictions on widespread solar pond development will be availability of land, water and salt. Large quantities of each are needed, especially to generate electricity because it takes 100 thermal kilowatts to produce a mere 10 electrical kilowatts.

The Salton Sea project, for example, will use 43 square miles of pond surface to generate 600 megawatts. On a more personal scale, a pond of one acre is needed to generate enough energy for 10 homes.

There are numerous sites around the country that appear virtually ready-made to serve as large solar ponds, including the Great Salt Lake and a number of bays and salt marshes. As for small ponds, they may soon be as common to farms and rural homes as silos and septic tanks.

"Controlled fusion can, in the long run...mean practically limitless energy, everywhere and forever."

The Fusion Alternative

Harold Seneker

Many engineers and energy specialists have been claiming that magnetic fusion could prove to be the ultimate solution to the world's energy crisis. Indeed, in 1980, the American government committed twenty billion dollars to research and development of fusion as an energy source. In the following viewpoint, Harold Seneker examines the potentials of controlled fusion and concludes that it is both an economical and desirable energy alternative.

Consider the following questions while reading:
1. **What does Mr. Seneker claim controlled fusion can do?**
2. **What type of engineering problem does the author admit still remains to be solved?**
3. **Why does Mr. Seneker conclude that fusion power would prove a "real blessing for mankind"?**

Harold Seneker, "Clean, Plentiful Energy on the Way." Reprinted by permission of *Forbes* Magazine from the November 24, 1980 issue.

As perhaps one of the last major acts of his Administration, President Carter signed the Magnetic Fusion Energy Engineering Act of 1980 on Oct. 7, committing the U.S. to spend $20 billion between now and the turn of the century. But what's a mere billion a year? Hardly worth counting these days. What really matters is that the new bill virtually hands over to private industry what may well be one of the most important technological developments in history. Controlled fusion can emasculate OPEC, drastically reduce the pollution that seems otherwise inevitably connected with industrial development and lessen the risk of war over scarce raw materials.

LIMITLESS ENERGY

It seems almost too good to be true, but controlled fusion can, in the long run—a generation from now—mean practically limitless energy, everywhere and forever...

Fusion is no longer just a theoretical possibility. Today it is an engineering problem. That's oversimplifying a bit; the engineering problems make the Manhattan Project look like Tinkertoys. But scientists already know half a dozen apparently feasible solutions. From here on, it's just a matter of time and money. Lots of both.

The key is "confinement." Hydrogen atoms must be forcibly brought together in a confined space at an unimaginably high temperature if they are to be fused together into helium and usefully throw off the prodigious energy thus released. In the H–bomb, an atomic explosion serves the purpose for the necessary billionth of a second; in the sun and other stars, enormous gravity does the trick; for steady, usable, energy production on earth, other methods are needed.

Magnetic confinement is the most advanced process: Atoms too hot to be contained by any solid wall can be trapped and insulated inside a magnetic field of the right shape, if it is strong enough. The Russians came up with the right shape—a hollow ring or, technically, a torus—and demonstrated it in the 1960s with a machine dubbed a *tokamak*. The Americans ran away with the concept, so it is a much–refined and enlarged tokamak that will be the heart of the $1 billion Fusion Engineering Device—a test facility intended to show by 1990 whether this approach is economically feasible...

FUSION ENGINEERING DEVICE

The question is how much it will cost. The Fusion Engineering Device is intended to answer that question for tokamaks by 1990. Not until then will anyone have a glimmering of whether fusion can be cost-comparable in the short run to all the other energy sources being worked on in and out of government: fossil fuels, synfuels, biomass, solar, wind, and so on. But note we say "in the short run." Long-run there can be no doubt that fusion, using cheap and plentiful sources for the energy it produces, will prove both economic and desirable.

Meanwhile, private industry is not wasting its time. General Atomic, a subsidiary of Gulf Oil, is running a fusion experiment in San Diego. Grumman and Ebasco Services are building a $400 million test machine for Princeton's Plasma Physics Laboratory. TRW has a small laser project going. McDonnell Douglas won the $70 million to $100 million contract to build and operate Elmo Bumpy Torus at Oak Ridge. The Department of Energy has also heard from Boeing, GE, Westinghouse and Bell Labs, all of whom are looking for ways to get into the game.

THE ULTIMATE ENERGY PRIZE

A fusion reaction yields energy more efficiently than fission, and generates far less of the radioactive garbage that has given fission a bad name... The great hope, as ever, is that some now unforeseen scientific breakthrough will yield the ultimate energy prize for America and the world.

Malcolm W. Browne, *Energy and the Way We Live,* 1980.

What attracts their interest most immediately is the Fusion Engineering Device, a $1 billion test facility.

Why Boeing or Bell Labs? Or Grumman? What do they have to do with atomic power? "We are looking," says John Clark, deputy director for magnetic fusion confinement at DOE, "for a parent organization to run this program — one that has a track record running other large programs. Its role will be coordinator — sort of a general contractor who will sub-contract big pieces of the work so we can spread knowledge of fusion technology through private industry."

MIT and Cal Tech (which operates the Jet Propulsion Laboratory) have such experience and have been nibbling, but Clark emphasizes it is corporations — none other than the good old military-industrial-complex corporation, in fact — that DOE is looking to.

TECHNOLOGICAL ADVANCES

DOE money is by no means the only motive for industry. Ask Stephen Dean, who was director of the department's magnetic fusion program in the 1970s, but left to set up a non-profit public interest association for this nascent industry. Explains Dean: "It gives them something for their very valuable teams of scientific personnel to do what's on the leading edge of technology, and get expertise in a whole list of new technologies peculiar to fusion." The new technologies, he explains, must be developed to make fusion work — super-conducting magnets, new heat-resistant metals and other materials, solutions to remote-materials handling and equipment-handling problems of unprecedented delicacy and difficulty, and the like — and these advances are very likely to yield profitable spinoff discoveries or commercial applications.

The fusion business has all the trappings already: Besides a trade association, it has its own connections to Congress...

Fusion even has its own political fringe group, the Fusion Energy Foundation, founded by one Lyndon H. LaRouche, Jr., also founder and chairman of the minuscule U.S. Labor Party. He and his disciples, who apparently revere a 19th-century physicist, Bernard Riemann, argue that, if society fails to develop new technology, it will die or regress into barbarism. The group publishes a magazine called *Fusion* and enthusiasts may be seen occasionally in airport terminals behind card tables, wearing three-piece suits and advocating nuclear power.

So 1980 may go down in history, not as the year Reagan beat Carter or the hostages were released, but the year in which fusion power entered the commercial stage. It would prove a real blessing for mankind. Fusion doesn't pollute; and it draws on heavy hydrogen isotopes, which exist in millions of years' supply in seawater, or can readily be made from lithium, one of the more common elements of the earth's crust. Cost aside, it may prove far more important to the world politically and militarily that no malevolent monopoly like OPEC nor phalanx of tanks can deny supplies of seawater or common rocks to any country.

DETERMINING PRIORITIES

Pretend that you are the U. S. Budget Director and the President has asked you for a recommendation as to how much money should be spent in the coming fiscal year on certain budget items. Assuming that you have a national budget of 240 million dollars, how much money would you advise the President to spend on each of the following items?

PART I

Instructions

Divide the 240 billion dollars among each of the items listed below. Be certain to assign the largest amounts of money to those items which are high on your list of national priorities and the smallest amounts to those which are low on your list.

$ _____ 1. National Defense

$ _____ 2. Foreign Aid

$ _____ 3. Space Research

$ _____ 4. Agriculture and Rural Development

$ _____ 5. Urban Development and Housing

$ _____ 6. Commerce and Transportation

$ _____ 7. Education

$ _____ 8. Health

$ _____ 9. Income Security: Social Security and Unemployment Compensation

$ _____ 10. Veterans' Benefits and Services

$ _____ 11. Pollution Control

$ _____ 12. Energy Research and Development

$ _____ 13. Select your own budget item

PART II

Now pretend that you are the Secretary of Energy and the U. S. Budget Director has allocated your department 10 billion dollars for energy development. How would you divide your available budget among each of the following energy sources?

Instructions

Divide the 10 billion dollars among the items listed below. Be certain to assign the largest amounts of money to those items which you believe to be the most promising energy sources for the future. Remember, that as Secretary of Energy, you can choose *not* to fund one or more of the items.

$ _____ 1. Coal

$ _____ 2. Synfuel development

$ _____ 3. Nuclear power

$ _____ 4. Solar Energy

$ _____ 5. Natural gas exploration

$ _____ 6. Oil exploration

$ _____ 7. Conservation education

$ _____ 8. Gasohol and bioconversion

$ _____ 9. Hydroelectric power

$ _____ 10. Geothermal energy

$ _____ 11. Select your own energy source

PART III

Instructions

STEP 1

The class should break into groups of four to six students.

STEP 2

Each group should choose a student to record its statements.

STEP 3

Each student should compare and defend his/her priorities as determined in Parts I and II with those of other members of the group.

STEP 4

The group recorder should poll the group to determine which priorities received the largest amount of money and which received the least.

STEP 5

Each group should compare its findings (from step 4) with those of other groups in the class.

BIBLIOGRAPHY

The following list of periodical articles deals with the subject matter of this chapter.

J.A. Briggs

Solar Power — For Real, **Forbes**, October 13, 1980, p. 142.

B.M. Casper *et al*

Real Energy Choice, **The Progressive**, February, 1980, p. 12.

Stephen Chapman

The Rebirth of King Coal, **The New Republic**, June 21, 1980, p. 15.

Todd Crowell

The Trouble with Fusion, **The Progressive**, April, 1980, p. 32.

L. Gasparello

Don't Curse the Darkness, Light a Candle, **Forbes**, May 12, 1980, p. 167.

Humanist

Alternative Energy Update, January/February, p. 61; March/April, p. 61; May/June, p. 53, 1980.

J. Mattill

Even Photovoltaics Have Social Costs, **Technology Review**, October, 1980, p. 80.

W.G. Reinhardt

Synfuels Other Side, **SciQuest**, January, 1981, p. 6.

Edward Teller

Shiva and the Politics of Laser Fusion, **Technology Review**, March, 1979, p. 64.

Time

Dangers of Counting on Coal, April 2, 1979, p. 66.

U.S. News & World Report

Special Report: Fuels for America's Future, August 13, 1979, p. 32.

Robert C. Willson

Synfuel, **Car and Driver Magazine**, August, 1980, p. 55.

APPENDIX OF ORGANIZATIONS

The editors have compiled the following list of private and public organizations concerned with energy issues.

American Gas Association
1515 Wilson Boulevard
Arlington, Virginia 22209
(703) 841-8400

An association of individuals, distributors and transporters of natural, manufactured and mixed gas which was founded in 1918. It publishes advertising and public information bulletins as well as publications of interest to its members.

American Nuclear Society
555 N. Kensington Avenue
LaGrange Park, Illinois 60525
(312) 352-6611

Founded in .1954, society members are physicists, chemists, engineers, educators etc. with professional experience in nuclear science or nuclear engineering. The society encourages research and provides a forum for the exchange of scientific and technical papers dealing with advancing the science and engineering of the atomic nucleus. The society publishes several monthly periodicals including *Nuclear News*.

American Petroleum Institute
2101 L Street, N.W.
Washington, D.C. 20037
(202) 457-7000

Institute members include producers, refiners, marketers and transporters of petroleum and allied products. The institute, founded in 1919, attempts to maintain cooperation between the government and industry, promotes foreign and domestic trade and the interests of the petroleum industry. It conducts fundamental research and provides extensive information services. The society publishes several hundred manuals and booklets and a *Statistical Bulletin*.

Alternative Energy Resources Organization
424 Stapleton Building
Billings, Montana 59101
(406) 259-1958

This organization, founded in 1974, attempts to encourage the transition from conventional to renewable energy sources. They promote renewable energy development through research, public education, community organizing and citizen representation. Their publications include energy activity books for children, fact sheets, pamphlets and the monthly *Sun-Times*.

Americans for Energy Independence
1629 K Street, N.W. Suite 1201
Washington, D.C. 20006
(202) 466-2105

This organization of individuals, founded in 1975, promotes the concept of "energy independence" derived from access to reasonably priced supplies and a primary reliance on domestic energy resources. The objective of the organization is to reduce U.S. demands for foreign oil imports, to accelerate research on new energy concepts and to encourage conservation. The organization publishes pamphlets and a quarterly newsletter.

Center for Energy Policy and Research
c/o New York Institute of Technology
Old Westbury, New York 11568
(516) 686-7578

The center was established in 1975 to disseminate information and conduct research into energy utilization and conservation, and to assist public and private organizations in the practical use of present and future findings in the energy field. The center provides monthly energy information to all the states and publishes monographs, reports, books and audio-visual materials.

Citizen Association for Sound Energy
P. O. Box 4123
Dallas, Texas 75208
(214) 946-9446

This association of individuals was founded in 1974 to promote effective action in the field of energy. They encourage conservative energy usage and research and are committed to renewable alternatives such as solar, wind, geothermal, hydrogen and coal gasification. Besides reprints and pamphlets, the association publishes a newsletter several times a year.

Consumer Information Center
Pueblo, Colorado 81009

A department of the General Services Administration which offers information of interest to consumers. Pamphlets available on energy include *Buying Solar*, $1.85 and *The Energy-Wise Home Buyer,* $2.00.

Department of Energy
Circulation
1E-218
Washington, D.C. 20585

The department publishes a free biweekly newsletter *Energy Insider.*

Fusion Energy Foundation
888 Seventh Avenue   Suite 2404
New York, New York 10019
(212) 265-3749

An organization of scientists, engineers and laypersons founded in 1974 to provide a forum for independent, scientific "discussion of fusion from the standpoint of comprehensive policy-making." The foundation publishes articles as well as the monthly *Fusion* and the quarterly *International Journal of Fusion Energy.*

Institute for Ecological Policies
9208 Christopher Street
Fairfax, Virginia 22031
(703) 691-1271

The purpose of the institute, founded in 1977, is to research and disseminate information about "environmentally sound and socially equitable policies" in areas such as energy. The institute collects energy consumption, conservation and renewable energy data for 3041 counties in the U.S. in order to coordinate a People's Energy Plan as an alternative to the National Energy Plan. They publish *People and Energy Magazine,* reports, books, a monthly newsletter *County Energy Plan,* and a bimonthly newsletter *Values Forum.*

National Energy Foundation
366 Madison Avenue   Suite 705
New York, New York 10017
(212) 697-2920

The foundation attempts to stimulate interest and to increase knowledge about the current energy situation among junior and senior high school students and teachers. Founded in 1976, the foundation sponsors "Student Exposition on Energy Resources," student seminars and teacher training programs. It maintains a central energy reference and resource center and publishes *Outlook,* a bimonthly newsletter, and *Energy Guide,* a quarterly teacher's guide.

Natural POWWER
6031 St. Clair Avenue
Cleveland, Ohio 44103
(216) 361-3156

An organization of businessmen, engineers, scientists and environmental activists, founded in 1975, which advocates replacement of toxic fuels and nuclear power with natural energy forces such as the sun, wind, tides and currents. (POWWER stands for Power of World Wide Energy Resources). They produce educational packets for elementary and secondary students and publish articles and a newsletter.

Nuclear Energy Women
7101 Wisconsin Avenue
Washington, D.C. 20014
(301) 654-9260

An organization of individuals in various energy industries and citizen advocacy groups which was founded in 1975. The group disseminates educational information to women's groups and others concerned about energy. The purpose of the group is to provide a clearer understanding of energy choices "with emphasis on providing the facts about nuclear power". They supply educational publications and audio-visuals on energy subjects.

Nuclear Information and Resource Service
1536 16th Street, N.W.
Washington, D.C. 20036
(202) 483-0045

An organization of anti-nuclear and safe energy alternatives groups and individuals founded in 1978. Its purpose is to assist persons and groups concerned about nuclear power, to provide information and advice to those attempting to halt nuclear plants and to promote alternatives to nuclear power. The organization produces information packets, distributes films and publishes *Groundswell,* a monthly journal, as well as pamphlets and fact sheets.

Solar Lobby
1001 Connecticut Avenue, N.W.   Fifth Floor
Washington, D.C. 20036
(202) 466-6350

An organization of individuals dedicated to the use of solar energy and conservation. It was founded in 1978. The group lobbies to encourage legislation favorable to the development of solar energy. They publish *Sun Times,* a monthly periodical.

Index

Khomeini, 46

LaRouche, Lyndon H., 134
lead time, 30-32, 107, 113
Libya, 36, 37, 82
Louisiana, 41
Lovins, Amory B., 62
Lovins, L. Hunter, 62

McCracken, Samuel, 16, 120
McElvay, Dr. Vincent, 40, 41, 44
meltdown, 57
Mexico, 37, 108, 123, 124
Missouri, 36
morality, 56
Murphy's Law, 72

Nader, Ralph, 57, 98
National Aeronautics and Space
   Administration (NASA), 127
National Petroleum Council, 117
natural gas, 21, 23, 24, 26, 29, 31,
   32, 36, 38, 40-44, 47, 58, 98, 110,
   115, 116, 121-123, 125
New Jersey, 43
Nigeria, 36, 37, 82
nuclear, 58, 59, 66, 68-75, 110, 124,
   129
nuclear power, 21, 31, 32, 47, 48,
   55, 56, 58-63, 65-67, 69-72, 74,
   75, 116, 119, 134
nuclear weapons, 61-65, 69, 75

Ohio, 128
oil companies, 32, 36, 38, 46, 80-87,
   89, 92-98, 100, 106, 108, 125
oil embargo, 19, 34, 36, 38, 46
oil sands, 109, 110, 112
Oklahoma, 36, 41
Oregon, 124
Organization of Petroleum Exporting Countries (OPEC), 15, 19-21,
   29, 31, 34, 44, 45, 49, 80-85, 87,
   97, 106-108, 113, 124, 132, 134

Pennsylvania, 59
Peralta, Manny, 110
petroleum, 24, 26, 35, 36, 41, 42, 97,
   107, 110, 112, 113, 115, 123, 124
photovoltaic cells, 115, 116, 121,
   122, 124
plutonium, 58, 64, 65, 72-74
political, 19, 21, 24, 46, 60, 93, 108,
   110, 113, 123, 125, 134
pollution, 47, 58, 62, 113, 124, 132,
   134
price controls, 26, 27
private enterprise, 34, 38
proliferation, 62, 63, 65, 73, 121
proved reserves, 23-27, 34

radioactivity, 56, 58, 59, 69, 70, 72-
   74, 122, 133
radiation, 56, 58, 59, 66-69
reactor, 56, 64, 65, 70, 72 121
regulations, 16, 23, 27, 37, 59, 89,
   96, 98, 113

renewable resources, 16, 62, 110,
   114, 115, 117
Riemann, Bernard, 134
Ross, Leonard, 62
Russia, 47, 69, 125, 132

salt domes, 37, 58
Salton Sea Project, 127-129
Saudi Arabia, 83, 93, 123, 124
Schlesigner, Energy Secretary, 85
scientific, 16, 28, 62, 112, 125, 133,
   134
security, 20, 28, 72, 74
shale oil, 35, 36, 38, 106, 109-112
Smith, Howard K., 67
social structure, 16, 71, 72, 74
soft energy, 16, 62, 120
solar collectors, 115, 116, 119
solar energy, 16, 21, 31, 32, 38, 46,
   72, 84, 105, 106, 108, 110, 114-
   124, 133
solar ponds, 126-130
South Africa, 106, 110
spot market, 87-89
Stobaugh, Robert, 19-21, 32
subsidies, 21
sulphur dioxide, 59
synthetic fuels, 21, 31, 38, 105-113,
   133

tar sands, 36, 38
taxation, 34
technology, 16, 24, 26, 31, 32, 38,
   63, 69, 70, 72, 74, 113, 117, 121,
   123, 126, 127, 129, 132-134
terrorism, 56, 58
Texas, 41
thermal energy, 127, 128
Three Mile Island, 62, 63, 67, 72,
   106
tides, 32
tokamak, 132, 133
torus, 132

unemployment, 21, 37
United Sates, 18-21, 23, 24, 26, 27,
   29, 40, 41, 46, 47, 63, 73, 87-89,
   93, 97, 98, 106, 108, 110-116, 124,
   132
U.S. Department of Energy, 32, 36,
   42, 87, 95, 113, 127, 129, 133, 134,
   140
U.S. Department of Justice, 87
U.S. Geological Survey, 23, 24, 41
Utah, 35

Venezuela, 36, 37

waste disposal, 56, 58, 72, 73
wind generators, 116
wind mills, 21, 32
Wright, Dick, 15, 16
Wyoming, 35

X-ray, 67

Yergin, Daniel, 19-21, 32

143

MEET
THE
EDITORS

Judy Smith received her BSE degree from the University of Arkansas, Fayetteville, with a combined major in English and journalism. She has supervised student publications at the high school level and is currently a freelance editor working on the Opposing Viewpoints Series.

Bruno Leone received his B.A. (Phi Kappa Phi) from Arizona State University and his M.A. in history from the University of Minnesota. A Woodrow Wilson Fellow (1967), he is currently an instructor at Minneapolis Community College, Minneapolis, Minnesota, where he has taught history, anthropology, and political science. In 1974-75, he was awarded a Fellowship by the National Endowment for the Humanities to research the intellectual origins of American Democracy.